Charles Morton Aikman

Milk - Its Nature and Composition

A Handbook on the Chemistry and Bacteriology of Milk, Butter and Cheese

Charles Morton Aikman

Milk - Its Nature and Composition

A Handbook on the Chemistry and Bacteriology of Milk, Butter and Cheese

ISBN/EAN: 9783337025199

Printed in Europe, USA, Canada, Australia, Japan

Cover: Foto ©berggeist007 / pixelio.de

More available books at **www.hansebooks.com**

MILK
ITS NATURE AND COMPOSITION

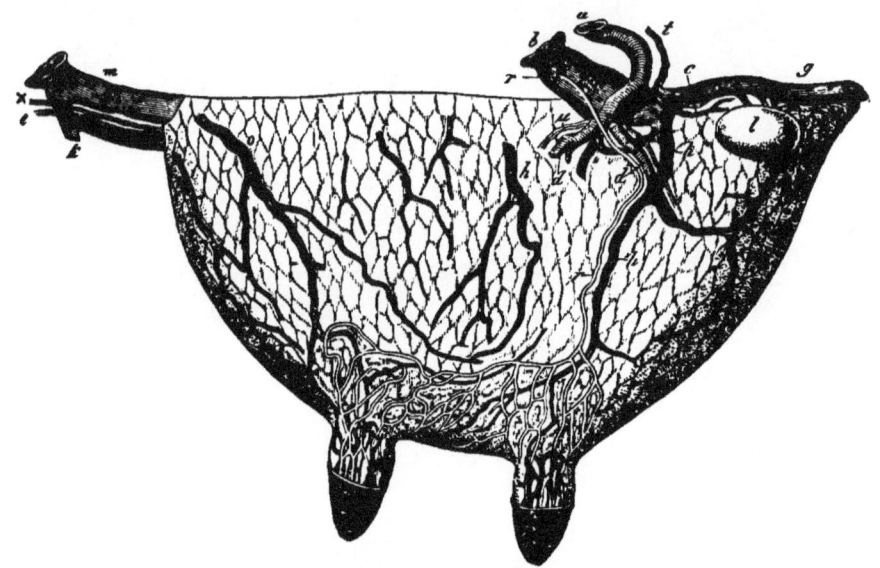

Cow's UDDER (after Fürstenberg).—The left side of the udder of a Dutch cow. The skin has been removed to show the course of the superficial arteries and veins, as also the lymphatic vessels and nerves. *a*, External pudic artery; *b*, external pudic vein; *c*, branch of artery supplying the posterior part of milk-gland, the lymphatic gland, etc.; *d*, posterior mammary artery, posterior artery of the milk gland; *e*, continuation of external pudic artery, ending in the skin and subcutaneous connective tissue near the breast bone; *g*, large vein from posterior part of mammary gland, from the lymphatic gland, the pudenda, etc.; *h*, posterior mammary vein; *l*, lymphatic gland situated on the posterior outer surface of the mammary gland; *m*, peritoneal or milk vein (vena subcutanea abdominis); *o*, anterior mammary vein; *r*, a lymphatic vessel receiving part of the superficial mammary lymphatics; *t*, mammary nerve formed by the union of the posterior branch of the iliac nerve (nervus ilio-hypogastricus) and a branch of the external spermatic nerve; *u*, posterior division of this nerve; *x*, continuation of *t*, ending in the skin at the navel.

MILK

ITS NATURE AND COMPOSITION

A HAND-BOOK ON THE

CHEMISTRY AND BACTERIOLOGY OF MILK,
BUTTER, AND CHEESE

BY

C. M. AIKMAN, M.A., D.Sc.

LONDON
ADAM AND CHARLES BLACK
1895

PREFACE

MUCH interest has been evoked during the last few years in dairy education. This has manifested itself in the institution of a number of dairy schools in various parts of the country. Unfortunately, the literature available to English readers on the science of dairying is still of a very limited and imperfect nature. Most of it deals rather with the art or practice of dairying than with the scientific side of the question.

The aim of the present volume is to give a short popular statement of the more important facts of the Chemistry and Bacteriology of Milk.

No attempt is made to deal with the practice of butter and cheese-making; but it is hoped that the scientific principles (so far as they are known) underlying these processes are stated in such a manner that they may be of assistance in explaining the operations of the dairy.

It is hoped that the work may possess interest for the general reader as well as for the farmer and student of agriculture, as milk is such a widely used article of diet, and exercises such an important influence on public health as a propagator of disease.

The science of dairying is largely dependent on the science of bacteriology, and this latter science is as yet in its infancy. As, however, the strides at present being made in the study of bacteriology are enormous,—as witness the fact that between the years 1890 and 1892 some 1,013 papers on the subject were published,—we may hope that the mystery still surrounding many problems in dairying will soon be dispelled. Bacteriology has already done much for the butter-maker in furnishing him with "pure cultures" of bacteria, and enabling him to secure a more uniform product than was hitherto possible. These pure cultures are not as yet used, so far as the Author is aware, in this country, but their use is now general in such countries as Denmark, Germany, Sweden, and Holland. It is much to be desired that our farmers should not be behind the foreigner in this matter.

Readers interested in the practice of dairying are referred to the English edition of Professor Fleischmann's elaborate treatise on *The Science and Practice of Dairying* (Blackie), translated and edited by Professor Wright and the Author.

The sources from which the Author has drawn are too numerous to admit of detailed acknowledgment. At the end of the book a short list of some of the books and treatises on the subject are given. He would, however, take this opportunity of specially acknowledging his indebtedness to the works of Fleischmann, Kirchner, Fürstenberg, Duclaux, Freudenreich, and Grotenfelt, and the numerous valuable Bulletins issued from time to time by the United States Government. Too great praise cannot be paid to the last-named Government for the enlightened policy which they pursue with regard to the circulation of valuable Bulletins, dealing with various aspects of agricultural science.

For the illustrations the Author is indebted to the writings of Dr. Fürstenberg, Professor Kirchner, Professor P. F. Frankland, Dr. James Bell, F.R.S., and Dr. Freudenreich.

ANALYTICAL LABORATORY,
128 WELLINGTON STREET, GLASGOW,
September 1895.

CONTENTS

CHAPTER I

The Structure of the Cow's Udder, and the Secretion of Milk

Definition of Milk—Structure of the Udder—The Formation of Milk . . . . Page 1

CHAPTER II

The Percentage Composition of Cows' Milk

Constituents of Milk—Percentage Composition of Milk—Variation in Composition of Milk . . . . . Page 9

CHAPTER III

The Constituents of Milk

Milk Fat—The Condition of the Fat Globules in Milk—Chemical Composition of Milk-Fat—The Albuminoids of Milk—Casein—Albumin—Lactoglobulin—Lactoprotein—Milk-Sugar—Ratio of the different Constituents of Milk—Mineral Constituents—Gases in Milk—Reaction of Milk—Sheep's, Goats', and Mares' Milk—Specific Gravity of Milk—Colostrum or First Milkings
Page 18

CHAPTER IV

Causes and Conditions Influencing the Quality and Quantity of Milk

Causes influencing Quantity of Milk Secreted — Difference in Morning and Evening Milk—Conditions influencing the Quality of Milk—Period of Lactation—Food—Influence of Excitement —Illness . . . . Page 42

CHAPTER V

The Changes which Milk Undergoes

Creaming of Milk—Souring—Amphoteric Reaction—Coagulation of Milk—Effect of Heat—Milk Faults—Means of preventing Changes in Milk—Preserved Milk . . . Page 53

CHAPTER VI

The Bacteria of Milk

Occurrence of Bacteria—Contamination of Milk by Micro-organisms —Importance of a Knowledge of Bacteriology to the Farmer— Description of the Different Micro-organisms — Method of Development of Bacteria—Size of Bacteria—Conditions influencing the Development of Bacteria—Number of Bacteria in Milk—Classification of Bacteria infesting Milk—Bacteria of a Useful Nature—Bacteria of an Indirectly Injurious Nature —Bacteria of a Directly Injurious Nature—Blue Milk Red Milk—Yellow Milk—Slimy or Ropy Milk—Bitter Milk— Lactic Fermentation of Milk—Butyric Fermentation—Casein Ferments — Alcoholic Fermentation — Kephir — Nature of Kephir Fermentation—Koumiss—Pathogenic Germs—Bacillus of Tuberculosis—Cholera Bacillus—Typhus Bacillus—Methods of Destroying or Regulating Bacterial Life in Milk--Sterilisation of Milk—Pasteurisation of Milk—Importance of Cleanliness in Handling Milk—Chemical Agents . . Page 63

CHAPTER VII

Butter—Importance of Bacteria for Butter-Making

Objects of Churning—Conditions influencing the Separation of Fat —Centrifugal Separators — Preliminary Souring of Milk or Cream before Churning—Conditions influencing Churning— The Bacteria of Butter — Aroma and Flavour of Butter — Number of Bacteria in Butter—Influence of Lactation, Breed, Age, and Food on the Quality of Butter—Chemical Composition of Butter—Margarine . . . . Page 114

CHAPTER VIII

Rennet and its Action

Occurrence of Rennet Ferment—Active Principle of Rennet—Coagulating Power of Rennet—Difference between Rennet-Curd and Acid-Curd—Action of Rennet—Forms in which Rennet is Used . . . . Page 137

CHAPTER IX

Cheese

Important Rôle of Bacteria in Cheese-Making—Conditions determining Quality of Cheese — Bacteria in Cheese — Chemical Changes Cheese undergoes during the Ripening Process— Cheese-Faults . . . . . . Page 148

CHAPTER X

Testing of Milk

Testing Fat by the Amount of Cream—Optical Methods of Determination of Fat—Determination of the Fat by Churning—

Determination of the Fat by the Addition of Reagents—Determination of the Specific Gravity—Formulae for Calculating the Composition of Milk from certain Data—Chemical Analysis of Milk . . . . . . Page 157

CHAPTER XI

Milk as a Food

Classification of Food Nutrients—Functions of Food—Digestibility of Food Nutrients—The Food Value of Milk—Digestibility of Milk—Effect of Boiling on Milk—Suitability of Milk as a Food—Value of Butter and Cheese as Food—Milk as a Food for Invalids . . . . Page 164

APPENDIX . . . . Page 175

INDEX . . . . Page 177

CHAPTER I

The Structure of the Cow's Udder, and the Secretion of Milk

Definition of Milk.—Milk may be described as the secretion of the mammary glands of the female mammal. It is a fluid which is secreted for a longer or shorter period after giving birth.

Before dealing with the nature, composition, and properties of milk, it will be well to say a word or two on the mode in which milk is secreted, and on the structure of the mammary glands. As we are here concerned almost entirely with cows' milk, it will be sufficient for our purpose to deal with the structure of the cow's udder.

Structure of the Udder.—The cow's udder, of which a diagram is given as the frontispiece of this book, consists of two milk-glands,—a right and a left,—which, in the adult cow, vary from about 9 to 12 inches in length, 6 to 12 inches in breadth, and 4 to

8 inches in depth (Fleischmann). These milk-glands are provided with outlet tubes, which are commonly known as the *teats*. As a rule, each milk-gland is provided with two teats, but sometimes with three; the third, when present, however, being in what is scientifically termed a "rudimentary form." The two glands are separated from one another by a

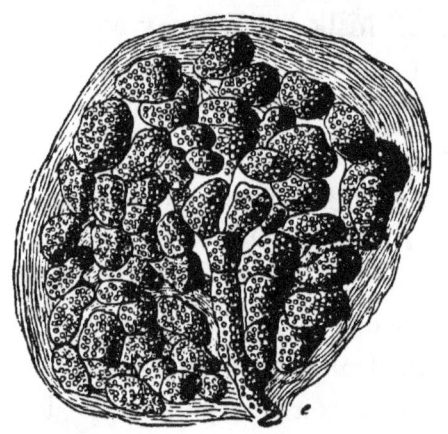

FIG. 1.—GLAND LOBULES. Magnified 60 times. *e*, Outlet tube.
(Fürstenberg.)

fibrous partition, running longitudinally, one of the chief functions of which is to support the udder. Each gland may be described as made up of a group of glandular structures, of different sizes, which are known as *lobes, lobules,* and *alveoli,* and each of which is smaller than the preceding one. The size of the alveoli varies from ·0047 to ·0078 of an inch in length, and from ·0035 to ·0043 of an inch in breadth (Fleischmann).

1]   STRUCTURE OF THE COW'S UDDER   3

The lobes, lobules (see Figs. 1 and 3), and alveoli (see Figs. 2 and 4) are provided with excretory ducts of corresponding size, which flow into the

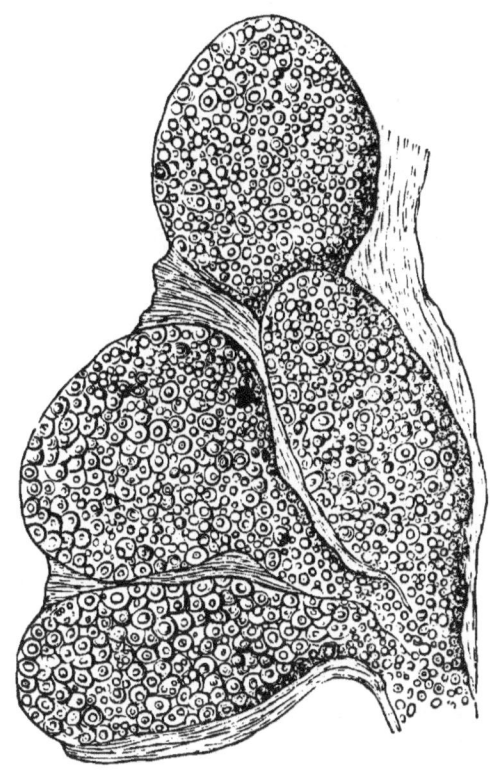

FIG. 2.—GLAND ALVEOLI. Magnified 200 times. (Fürstenberg.)

so-called *milk-cisterns* or *milk-reservoirs*, which are situated immediately above the teats. When a duct, therefore, is traced into the gland, it will be found to become sub-divided into smaller ducts, and these, in their turn, into still smaller ones; while round the smallest of the ductlets are clustered several alveoli. Of these milk-cisterns there are four, one above each teat. They thus act, as their name indicates, as cisterns for supplying the teats with milk.

The **alveoli** or pouches above mentioned are surrounded by a structureless membrane (*tunica propria*), the internal surface of which is thickly covered with

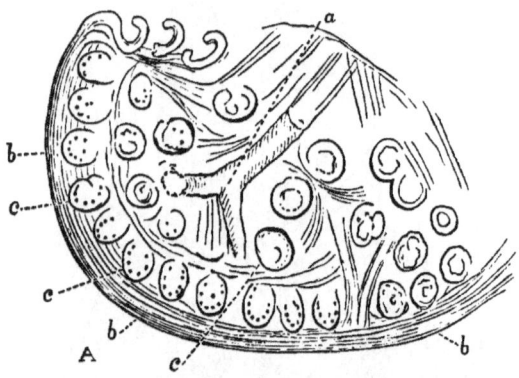

FIG. 3.—HALF DIAGRAMMATIC VIEW OF A SECTION THROUGH A LOBULE OF THE MAMMARY GLAND (after Klein, *Atlas of Histology*, Plate 40, Fig. 1). Magnified 45 diameters. *a*, A duct dividing into two branches; *b, b, b*, connective tissue surrounding and going between the ultimate pouches of the gland; *c, c, c*, the pouches or *alveoli* of the gland, the dots representing the cells lining them.

epithelial cells (see Figs. 5 and 6). The membrane is further covered with a system of capillary blood-vessels (see Fig. 7), which brings the blood near the cells, and thus nourishes them. Each alveolus possesses a minute cavity in its centre (see Fig. 4), in which the cells or their products can accumulate. When milk is not being formed the cells possess a granular appearance, and the central cavity above referred to is small; but during the process of secretion the cavity becomes dilated.

The **teats** consist of two parts: the upper part, which is known as the *basis*, and the lower, which is known as the *nipple*. The nipple of the teat con-

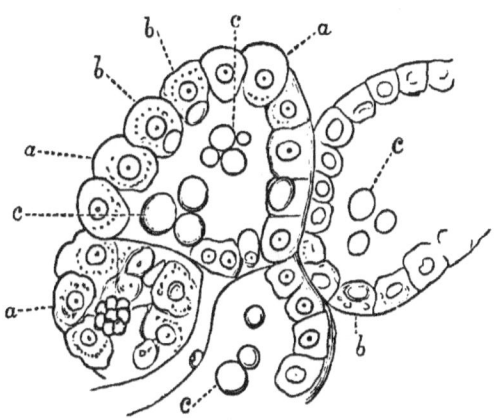

Fig. 4.—A portion of the mammary gland, magnified about 400 diameters, showing one complete and two incomplete alveoli. *a, a, a*, Short, columnar epithelial cells lining the alveolus, each having an oval or rounded nucleus; *b, b, b*, epithelial cells containing next the interior of the alveolus a milk globule; *c, c, c,* milk globules which have been set free from epithelial cells.

tains numerous blood-vessels, and has at the base muscular fibres (see Frontispiece).

It has been calculated that the two milk-glands and the four milk-cisterns in the udder of an average milk cow, when the udder is not in any way distended,—that is, as it would be after the cow has been milked,—have a cubic capacity of from $10\frac{1}{2}$ to $11\frac{1}{4}$ pints.

The Formation of Milk.—The exact method in which milk is formed in the udder is, as yet, far from having been clearly demonstrated. Two important theories have been advanced. According to the older one, which held sway in scientific circles up to the year 1840, milk is formed directly from the blood—it is, in fact, a sort of filtered blood. Accord-

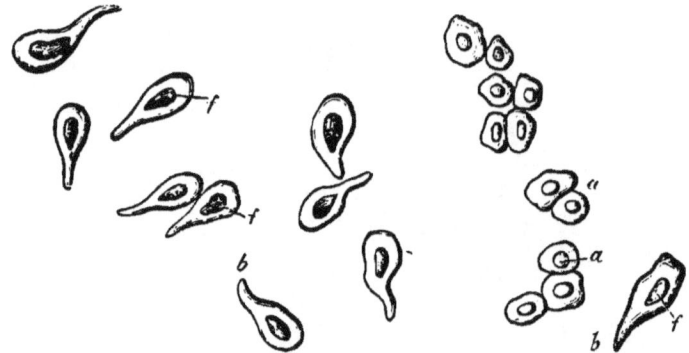

FIG. 5.—UNCONNECTED EPITHELIAL CELLS. Magnified 600 times. *b*, Protuberance resting on basement membrane; *f*, nucleus. (Fürstenberg.)

FIG. 6.—EPITHELIAL CELLS. Magnified 600 times. *a*, Attached together; *b*, free cell. (Fürstenberg.)

ing to this theory, the milk-glands merely act as a filter. The other theory regards milk as the product of the decomposition of the epithelial cells. There can be no doubt which of these two theories commands most support from the data available. In the first place, if milk be merely filtered blood, its quantity and quality will be directly influenced by the food of the cow. That this is not so, practical men have long known. No doubt food has a certain indirect influence on the quality of milk, as we shall point out farther on; so far, however, as we at present understand the subject, the direct influence of food, as has been again and again proved,—as, for example, by changing the diet,—is insignificant in extent. That milk is not filtered blood is further supported by the fact that none of the organic constituents of milk are present in the blood. Again, we find the proportion of the different ash constituents in blood and in milk to be different. In the former it is found that sodium salts predominate; while in the latter potassium salts are more abundant. On the other hand, considerable support to the correctness of the second theory seems to be afforded by the fact that in *colostrum*, the name applied to the milk yielded immediately after giving birth, are to be found certain minute corpuscles, the so-called *colostrum corpuscles* (*corps granuleux*), which exhibit traces of cell structure.

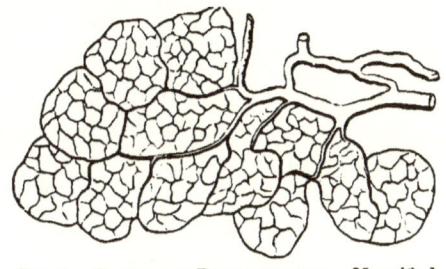

Fig. 7.—Capillary Blood-vessels. Magnified 180 times. (Fürstenberg.)

How far, however, even this second theory satisfactorily accounts for the formation of milk is doubtful. It may be, indeed, that the various constituents of the blood and the lymph bodies, as well as the substance of the gland cells, are all utilised in the formation of the constituents of milk. With regard to the fat of milk, at any rate, it seems that it is partly formed from the fat of the blood and partly from the cell substance.

CHAPTER II

The Percentage Composition of Milk

THE composition of the milk of different animals is practically the same, although a considerable variation occurs in the proportion in which its different constituents are present.

Constituents of Milk.—The chief constituents are *water, fat, protein,* or albuminoid matter (of which *casein* is by far the most abundant), *milk-sugar,* and *mineral matter.* The variation in the percentage of these constituents may, under exceptional circumstances, be considerable.

The following table, after König, gives the average composition of the milk of different animals :—

	Water.	Casein.	Albumin.	Ash.	Milk-sugar.	Fat.
Human .	87·41	1·03	1·26	0·31	6·21	3·78
Cow .	87·17	3·02	0·53	0·71	4·88	3·69
Ewe .	80·82	4·97	1·55	0·89	4·91	6·86
Goat .	85·71	3·20	1·09	0·76	4·46	4·78
Mare .	90·78	1·24	0·07	0·35	5·67	1·21
Ass .	89·64	0·67	1·55	0·51	5·99	1·64
Sow .	84·04	7·23		1·05	3·13	4·55
Bitch .	75·44	6·10	5·05	0·73	3·08	9·57
Buffalo .	81·41	5·85	0·25	0·87	4·15	7·47
Camel .	86·57	4·00		0·77	5·59	3·07
Cat .	81·63	3·12	5·96	0·58	4·91	3·33
Mule .	91·50	1·64		0·38	4·80	1·59
Lama .	86·55	3·00	0·90	0·80	5·60	3·15
Elephant	67·85	3·09		0·65	8·84	19·57
Porpoise	41·11	11·19		0·57	1·33	45·80

From the above it will be seen that a considerable variation in the composition of the milk of different animals exists, the milk of the porpoise being the richest in solids, while that of the elephant comes next; whereas that of the mule contains the largest percentage of water. The most wide variation, it will be seen, is in the fat, the other constituents being present in comparatively uniform amounts.

As we have said, considerable variation in the composition of the milk of the same species of animal may exist, and this is the case with regard to cows' milk, with the composition of which we are most familiar.

At first sight, therefore, it would seem rather difficult to say exactly what amount of these different constituents constitutes normal cows' milk.

The following tables, drawn up by different authorities, illustrate this variation, and give, at the same time, what is considered by their respective authors to be the composition of average milk:—

Percentage Composition of Milk

	Fleischmann.		Kirchner.		American.[1]		English.
	Average.	Limits of Variation.	Average.	Limits of Variation.	Average.	Limits of Variation.	English average of 120,540 Samples of Milk (Vieth).
Water	87·75	87·5 to 89·5	87·5	83 to 90	87·00	81·1 to 91·4	87·1
Fat	3·40	2·7 to 4·3	3·4	·8 to 8	4·00	2 to 8	4·1
Nitrogenous substances, commonly known as Casein or Caseous Matter	3·50	3 to 4	3·9	2·28 to 5·73	3·30	2 to 4·5	8·8
Milk-sugar	4·60	3·6 to 5·5	4·5	3 to 6	4·95	4 to 5·5	
Ash	·75	·6 to ·9	·7	·6 to ·9	·75	·6 to ·9	
	100·00		100·00		100·00		
Total Solids	12·25	9·9 to 14·7	12·5	7·2 to 12·63	13·00	8·6 to 18·9	12·9

(Solids not Fat bracket spans Nitrogenous substances, Milk-sugar, and Ash rows.)

[1] Results of all complete American analyses of milk, up to 1891 (see *American Experiment Station Record*, vol. v. No. 10).

With reference to the above figures, it may be added that, as far as one ingredient is concerned, viz. the fat, limits even wider than those above cited have been published. Thus a sample of milk has been found to contain only ·25 per cent [1] of fat; while another sample has been found to contain as much as 11·06 per cent.[2] These are respectively the lowest and highest percentages of fat found in samples of genuine milk, so far as the author is aware, that have yet been published.

The results above quoted may be summarised as follows—the average composition being obtained by averaging the results :—

COMPOSITION OF MILK.

	Average.	Limits.
Water	87·34	81 to 91
Total Solids	12·66	7·2 to 18·9
	100·00	
Fat	3·72	·25 to 11·0
"Solids not Fat" *	8·93	5·6 to 12·6

* Consisting of—

Casein and Albumin	3·594
Milk-sugar	4·614
Ash	·730

Variation in Composition of Milk.—With respect to the variation in the amounts of the different constituents of milk, it is not to be inferred

[1] See *Milch-Zeitung* 22 (1893), p. 804.
[2] See *The Analyst* (1893), pp. 1-12.

that such wide limits, as have been just indicated, actually occur under normal circumstances. Indeed, it need scarcely be pointed out that, if this were the case, detection of the adulteration of milk would be rendered a well-nigh impossible task. Where any wide variation from what is above stated as the average composition of milk is found, the conditions under which such milk is obtained may be safely assumed to be abnormal; for, under normal conditions, the composition of milk is comparatively uniform, with the exception, perhaps, of the amount of fat. Furthermore, it must be borne in mind that this uniformity of composition is rendered more probable by the fact that milk sold for consumption is generally "dairy" milk—that is, the mixed milk product of several cows. Wide variation in the composition of such milk is not likely ever to occur. In such a case, even supposing the milk of any single cow should exhibit an abnormal composition, the fact that it is mixed with the milk of a number of other cows, several of which may yield milk richer in quality than average milk, serves to a large extent to counteract the influence of the abnormal milk in lowering the quality.

In forming an opinion as to the genuineness of a sample of milk, the more important data to be taken into account are the "total solids," the fat, and the "solids not fat"; and the Society of Public Analysts has drawn up, as a result of a thorough consideration of the whole subject, the following standards, which express the minimum amount in which these ingredients should be present in a genuine sample of milk :—

Total Solids	11·5	per cent
Fat	3·0	,,
Solids not Fat	8·5	,,

In support of the reasonableness of the above standards, it may be mentioned that Sir Charles Cameron, M.D., the Public Analyst for Dublin, has never found the total solids in mixed milk to fall below 12 per cent. The practical importance of the point is very great, since on it depends the efficient working of that portion of the Sale of Foods and Drugs Act which refers to the adulteration of milk : a form of adulteration which, we may add, is most commonly practised. For example, if a sample of milk should be found to contain only 2 per cent of fat, on what grounds, it may be asked, would a Public Analyst be justified in regarding it as adulterated, since the analyses of milk known to be genuine have shown, as we have above pointed out, an amount of fat less than this ? A superficial consideration of the subject seems to point to the conclusion that the condemning of such a sample would be unjustifiable. Now, as the law at present stands, there is no doubt that there is something to be said for such a contention.

Genuine milk containing such a low percentage of fat is no doubt of very rare occurrence, yet it may occur; and not to take this into consideration may seem to savour of injustice. But the question arises: Ought it to make much difference, where the quality of the milk is so poor, whether the sample has been adulterated in the technical sense—that is, been submitted to a watering process subsequent to its being obtained from the cow's udder—or not? for it is obviously unfair that a person who expects to obtain ordinary milk should be provided with abnormal milk. A man selling such abnormal milk should be treated very much in the same way as a man selling adulterated milk. Milk, therefore, should be defined under the Sale of Foods and Drugs Act as the *normal* secretion of the mammary glands of the cow. How far this is the interpretation of the existing law on the subject is a question open to opinion. Certainly, if this were more widely recognised, many of the vexatious disputes which the sanitary authorities too often experience in their attempts to obtain convictions in the case of undoubtedly adulterated samples of milk would be done away with. A satisfactory solution of the difficulty would be effected by selling milk on the basis of its analysis; and this, there can be little doubt, while accompanied by certain difficulties, will be the method sooner or later adopted.

The variation in the composition of the milk of different cows is due to a number of causes: among them may be mentioned the *individuality* and *breed* of the cow, the *period of lactation*, and the *feeding*. Before, however, considering the conditions which influence the quantity as well as the quality of milk, a short description of its different constituents may be given.

Before concluding the Chapter, a list of some of the standards adopted in some of the States of America may be given. Milk showing a poorer composition than the standard is regarded as adulterated.

PERCENTAGE COMPOSITION OF MILK

State, City, etc.	Percentage by Weight of Solids.		
	Solids not Fat.	Fat.	Total Solids.
New York. Law 1893	9·00	3·00	12·00
Maine. Law 1893	9·00	3·00	12·00
Michigan. Law 1889	9·50	3·00	12·50
Iowa. Law 1892	...	3·00	...
New Hampshire. Law 1883	...	...	13·00
Ohio. Law 1889	9·38	3·12	12·50
Oregon. Law 1893	...	3·00	...
Vermont. Law 1888	9·25	3·25	12·50
Wisconsin. Law 1889	...	3·00	...
Pennsylvania. Law 1885	9·50	3·00	12·50
New York. Law 1884	9·00	3·00	12·00
New Jersey. Law 1882	9·00	3·00	12·00
Massachusetts. Law 1886	9·30	3·70	13·00
May and June	...	...	12·00
Minnesota. Law 1889	9·50	3·50	13·00
Columbus, Ohio	9·37	3·12	12·50
Baltimore, Md.	...	...	12·00
Denver, Col.	...	...	12·00
Lansing, Mich.	...	3·00	12·50
Madison, Wis.	...	3·00	...
Des Moines, Iowa	...	3·50	13·13
Portland, Oregon	...	...	12·00
Omaha, Nebraska	...	3·00	12·00
U.S. Treasury Department	9·50	3·50	13·00
Philadelphia, 1890 Ordinance	8·50	3·50	12·00

CHAPTER III

The Constituents of Milk

THE constituents of milk are *fat, casein, albumin, lactoglobulin, lactoprotein, milk-sugar, mineral matter,* certain *gases* in solution, as well as minute traces of other bodies.

Milk-Fat.—The fat may be considered as the most important constituent of milk, not only because its commercial value, as well as the commercial value of one of its great products,—cheese,—depend on it, but because it is the chief constituent of the other and still more important milk product, butter.

The colour and opacity of milk are largely due to the presence of its fat. It exists in the milk in the form of minute globules,—known as the *fatty* or *milk globules,*—so minute as to be quite invisible to the naked eye. They are apparent, however, if we submit a drop of milk to microscopic examination. It will be seen from the accompanying diagram (see Fig. 8) that they differ very considerably in their size, the largest being about 6¼ times the size of the smallest. Careful investigations carried out by Fleischmann have shown that the diameter of the former measures ·00039 inch, and that of the latter ·0000624 inch. Between these two sizes we have

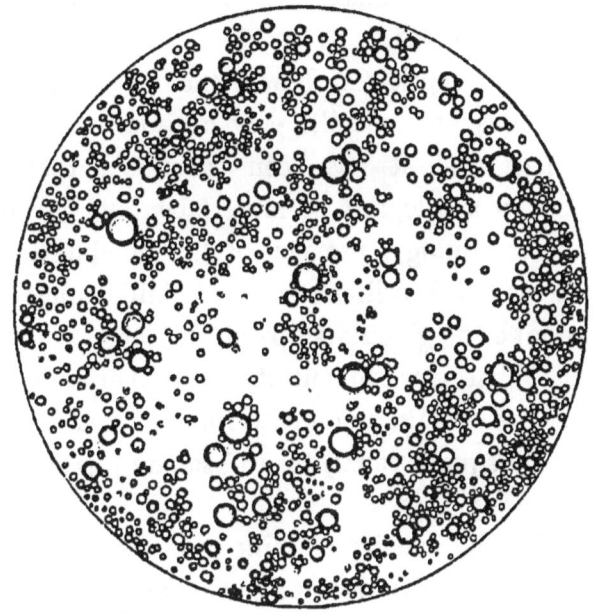

FIG. 8.—MICROSCOPIC APPEARANCE OF MILK, containing 3·6 per cent of fat. (Kirchner.)

a large variety of globules. It would seem as if some definite ratio existed between their size and their number, the smaller ones being more abundant than the larger ones. Indeed it seems probable that the weight of all the globules of one size is equal to the weight of all the globules of another size; or, to put it in another way, the number of globules of different sizes is in inverse ratio to their size. It is also to be noted that the size of the fatty globules in the milk of the same cow is not at all times, and under all conditions, the same ; and probably even a greater difference exists in those of the milk of cows of different breeds.

Some conception of how numerous these minute fat globules are in milk may be inferred from the statement that it has been calculated that their number in a kilogram of milk (about $2\frac{1}{4}$ lbs. or $1\frac{3}{4}$ pints), containing 4 per cent of fat, amounts to from eighty thousand millions to twenty billions, according as we take the globule to be of the largest or of the smallest size; or, to put it in another way, their surface area would amount to from 25 square metres (about 30 square yards) to 157 square metres (about 188 square yards).

While the opacity of milk is chiefly due to the fatty globules, it is also due to the fact that a portion of the mineral and nitrogenous matter is present in a state of suspension. Milk in thin layers, however, is not absolutely opaque, and it has been proposed to test the percentage of its fat by taking advantage of this fact and estimating its degree of opacity. A thin layer of milk is examined, and the amount of light it retards is noted. But this so-called "optical method" of estimating fat is very unreliable, since it has been found that the same weight of fat retards more light when it is in the form of small globules than when it is in the form of larger globules; so that two samples of milk, containing equal amounts of fat, but the fat in the one in the form of larger globules than in the other, might give different results when thus tested.

The Condition of the Fat Globules in Milk.—A point of the very highest interest, and one which has engaged considerable attention, is the condition in which these fat globules are present in milk. It was till comparatively recently believed that they were surrounded by a solid albuminous membrane which became ruptured when milk was churned. This theory seemed to derive considerable support from the behaviour of the fat under different circumstances. For one thing, it was found almost impossible to obtain fat from milk absolutely devoid of nitrogenous matter. Another reason for believing in the existence of a membrane was the isolation which the fatty globules maintained in the milk. If there is no albuminous envelope, why do the globules not coalesce? it may be asked. A further striking anomaly seemed also to point to the existence of a membrane, viz. that the fat is in a liquid condition, although its temperature may be far below its melting-point.[1]

---

[1] A proof of the liquid condition of the fat in the globules at even fairly low temperatures, viz. those approaching the freezing-

But all these seemingly plausible arguments in favour of the existence of a membrane can be easily explained on grounds which do not require the assumption of a membrane. In the first place, the difficulty of obtaining milk-fat free from albuminous matter is due simply to the fact that minute quantities of these bodies cling to the fatty globules, and, although not forming any solid membrane, may yet form a thin liquid envelope. Again, the isolation of the fatty globules, although not so easily explained, does not need the supposition of a membrane. It may be explained by the semi-solid condition of the casein and of a part of the mineral matter. According to Babcock, milk, after being kept for a short time, has formed in it a body resembling blood-fibrin. This body has the effect of rendering the milk less limpid, and at the same time helping to prevent coalescence of the globules. A further argument in support of the membrane theory is based on the behaviour of milk towards ether, which, it is well known, is an excellent fat solvent. If milk be shaken up with ether, it is found that the ether does not dissolve out the fat globules, as is proved by the fact that there is no diminution in the opacity

point of water, is afforded by submitting them to microscopic examination, when it will be seen that they maintain their globular form. On solidifying them at a considerably lower temperature, however, they will be seen to lose their globular form and assume different shapes, due to solidification.

of the milk. If, however, in addition to ether, a potash or soda solution be added, the ether speedily dissolves the fat, and the milk becomes much clarified. According to the supporters of the membrane theory, the alkaline solution dissolves the membrane and permits the ether to act upon the fat. That this, however, is not the true explanation has been strikingly proved by Soxhlet, who found that milk to which potash or soda solution had been added is not clarified by the addition of benzene or chloroform—equally powerful fat solvents. It seems clear that the peculiar action of ether is due to another cause, which need not here be discussed. Lastly, the liquid state of the fat in the globules may be explained as due simply to their isolation, and furnishes an example of what is chemically known as *superfusion*. It is a well-known fact that any liquid, if it is in a sufficiently fine state of division, may be cooled down below its freezing-point without becoming frozen. A sudden impact, however, will convert it into the frozen state.

If, then, it may be said that the fatty globules have an envelope, it can amount to nothing more than a mere liquid envelope, such as the film of a soap-bubble.

Chemical Composition of Milk-Fat.—Milk-fat differs from all other known kinds of fat in its complex and highly characteristic chemical composition, and may be said to be characterised among fats by its fine flavour. There can be no doubt, also, that the fine state of its division in milk or cream makes it, when taken in either form, more easily digested than other fats. On the other hand, it is more susceptible to decomposition than they are.

Fats are bodies made up of *glycerides*—organic bodies, which present in their composition a certain resemblance to the salts of inorganic chemistry. These glycerides are composed of *glycerin* and a *fatty acid*.[1] Milk-fat, like most other fats, is chiefly made up of three of these glycerides, viz. *stearin, palmitin,* and *olein,* which together constitute about 91 per cent of its composition. But what is especially characteristic of milk-fat is the presence of some seven other fats, viz. *butyrin, capronin, caprylin, caprin, laurin, myristin,* and *butin.* Of these, butyrin and capronin predominate in amount, the other five being present in very small quantities. The melting-point of these different glycerides varies; that of stearin being 55° C. (123° Fahr.), palmitin 62·8° C. (144·6° Fahr.), and myristin 31° C. (88° Fahr.). All the other fats, with the exception of butin and caprin, which are present in very minute quantities, and which, therefore, have little effect in influencing the

[1] In butter-fat glycerin may be said to be present to the amount of 4·5 per cent, while the fatty acids are present to the extent of 94·5 per cent.

melting-point of butter, are liquid at ordinary temperatures. Olein, indeed, only becomes solid at the freezing-point of water. Since the proportion in which the nine fats are present in milk-fat is a variable one, its melting-point is also variable, and may be stated at between 29° and 41° C. (84° to 106° Fahr.) The nature of the fat is largely determined, however, by the relative percentage of stearin, palmitin, and olein.

The fatty acids which go to form the above fats may be divided into two classes, viz. those which are insoluble and non-volatile—and this class is by far the largest in amount, and includes palmitic, stearic, oleic, butic, and myristic acids—and those which are soluble and volatile. The respective amounts of these two classes of fatty acids in milk-fat, as we have just indicated, varies at different times. Among the conditions which influence their amount is the stage of lactation, the age, breed, and feeding of the cow. The percentage of volatile fatty acids was found by Duclaux in nineteen samples of butter-fat to vary between 5·77 and 7·95 per cent, the average being 7·3. They consisted almost entirely of butyric (4·58 per cent) and capronic (2·7 per cent) acids. The presence of these volatile acids is of importance, as it affects the quality of the milk-fat and influences its flavour. It may be said that the higher the percentage of fatty acids in the milk-fat, the better the quality of the butter. It has been thought that the composition of the fatty globules of different sizes in milk varies, the fat in the larger globules being of finer flavour than the fat in the smaller globules (see Butter, Chapter VII., p. 114). In experiments carried out at Göttingen by Goetz, the influence of breed on the size of the fat globules has been investigated. Thus it was found that in the milk of a Jersey cow their size varied between ·009 and ·0042 millimetres; in that of an East Friesian cow (German breed) from ·0063 to ·0021 millimetres; and in that of a Simmenthal cow (German breed) from ·004 to ·0027 millimetres. With respect to the influence of food in causing a variation in the percentage of the fatty acids, a number of experiments have been carried out,[1] which go to show that this is considerable; while with regard to the influence of lactation, it has been found that the maximum percentage of fatty acids in the milk-fat occurs in from five to seven days after calving.

It may be added, in conclusion, that the colour of the milk-fat varies from white to yellow. If pure milk-fat be kept from the access of air and light for some time it becomes rancid. This is caused by the decomposition of the bodies out of which it is formed, with the result that small quantities of fatty acids, particularly butyric acid, are set free. When

[1] By Mayer, Kirchner, and Ladd.

freely exposed to the air, and in the presence of sunlight, this decomposition goes on more quickly. It may be here pointed out that not merely are certain fatty acids set free, but actually new acids, such as formic acid, are formed by the absorption of oxygen. In this process of decomposition the fat assumes a white colour. The conditions which influence the percentage of fat in milk will be discussed in the next Chapter, under the wider question of the conditions which influence the composition of milk.

The Albuminoids of Milk.—The albuminoids of milk, from a chemical point of view, are the most interesting of the milk constituents. More discussion has taken place with regard to their nature than with regard to the nature of all the other constituents of milk. Some, like the eminent French investigator Duclaux, have held that there is only one albuminoid substance in milk. This theory, however, seems to be at variance with the behaviour of milk when treated with certain reagents; and there can be little doubt that there are several albuminoids in milk. Of these, *casein*, or what has been called *caseous matter*, or *caseinogen*, to distinguish it from pure casein, is the most abundant, and forms about 80 per cent of the total nitrogenous matter. The other albuminoids are *albumin*, which is next abundant in amount to casein; *lactoprotein*, which has also been called albuminose or galactine; and *lacto-globulin*.

Casein.—According to Kirchner, casein (*i.e.* casein proper) varies from 2 to 4·5 per cent, and is, on an average, present in milk to the extent of 3·2 per cent. Its percentage composition is as follows :—

Carbon	53·00
Hydrogen	7·12
Nitrogen	15·65
Oxygen	22·60
Sulphur	·78
Phosphorus	·85
	100·00

From the above it will be seen that milk casein contains both sulphur and phosphorus. Casein is characterised by the presence in it of a substance called *nuclein*, a body not found in albumin. What is usually known as the casein in milk is not, as we already mentioned, pure casein, which is a body insoluble in water. There seems to be no doubt that the casein, in milk, is combined in some way with lime. This compound, which, as we have said, is sometimes called caseous matter or caseinogen, differs from pure casein by being more soluble. It is not, however, strictly speaking, soluble in water, but forms with it a bulky, colloidal (or glue-like) mass. This is proved by the fact that if milk be filtered through porous clay the caseous matter does not pass through, like the other albuminoids, which are in true solution. The presence of this glue-like caseous matter in milk exerts an important influence on the state of the fatty globules, as we shall see later on, by preventing the more minute ones from separating out from the main body of the milk by rising to its surface. The caseous matter of milk may be regarded, then, as being of the nature of a salt, in which the casein proper takes the part of the acid, and the lime that of the base. The proportion in which the casein and lime are respectively present are 100 parts of the former to 1·55 of the latter. As we shall see farther on, the action of rennet or dilute acids is to decompose this caseous matter, and to precipitate it from its colloidal form into an insoluble one.

The caseous matter in its semi-dissolved condition possesses a certain amount of opacity. The opacity of milk is, therefore, due to its caseous matter as well as to its fatty globules. This is seen from the fact that creamed milk, which contains as little as ·1 per cent of fat, presents considerable opacity. It is this body which forms the chief constituent of the curd or coagulum of milk, the nature of which, according to the conditions under which it is produced, will be discussed when dealing with cheese.

It has been noticed that, when milk is allowed to stand for some time, an alteration of the casein into peptones may take place. Thus, a German investigator has found that, in milk standing eight hours, a loss of caseous matter amounting to ·25 per cent (that is, 10 per cent of the original quantity) took place, owing to such conversion. A consideration of this fact points to the importance of coagulating milk destined for the manufacture of cheese as speedily as possible.

Albumin.—The albumin of milk does not seem to be identical with blood or serum albumin, and hence it has been called *lactalbumin*. Its composition (Sébelien) is as follows :—

Carbon	52·19
Hydrogen	7·18
Nitrogen	15·77
Oxygen	23·13
Sulphur	1·73
	100·00

Albumin differs from casein in not containing phosphorus, and from the other albuminoids of milk in containing sulphur. Its average percentage in milk, according to Kirchner, is ·6; its amount varying from ·2 to ·8 per cent. It is, as has already been mentioned, in a state of solution in milk. It is soluble in dilute acids and rennet, and is not coagulated along with the casein in cheese-making. It thus chiefly passes into the whey under such circumstances. By heating the milk to from

70° to 75° C. (158° to 167° Fahr.), however, it is coagulated. While present in normal milk in but small quantities, its amount in colostrum is very much greater.

Lactoglobulin.—This body is present in very minute traces in ordinary milk (according to Sébelien only a few parts per million) ; but in colostrum it may amount to as much as 8 per cent. Like albumin, it is in a state of solution, and can be precipitated from milk after the caseous matter has been removed by treating it with magnesium sulphate. When the milk is heated to a temperature of from 70° to 76° C. (158° to 169° Fahr.) it is coagulated.

Lactoprotein.—If milk be treated with acetic acid to precipitate the casein, then boiled to remove the albumin and lactoglobulin, it will be found that there is still another albuminoid substance left. This body, which is present to the extent of from ·08 to ·19 per cent—on an average ·13 per cent—in fresh milk, belongs to the peptone group of substances. It has been variously named *lactoprotein, albuminose,* and *galactine.* According to Babcock, there is yet another albuminoid present in milk, viz. *fibrin,* which amounts to ·1 per cent.

Milk-Sugar.—Milk-sugar is characteristic of milk, since it does not occur elsewhere. It is very easily decomposed when in a state of solution,—as it is in milk,—the result being its conversion into lactic acid. This change is effected by bacteria. The class of bacteria effecting it are of common occurrence, and have been found in great numbers on the cow's udder, in the byre, in milk-vessels, etc. They thus obtain easy access to the milk, and the souring of milk through their action follows sooner or later. The following is the composition of milk-sugar :—

Carbon	40·00
Hydrogen	6·10
Oxygen	48·90
Water	5·00
	100·00

Its chemical formula is $C_{12}H_{22}O_{11}$, H_2O. The conversion of milk-sugar into lactic acid may be explained by the following equation, in which one molecule of milk-sugar is converted into four molecules of lactic acid :—

$$C_{12}H_{22}O_{11}H_2O = 4(C_3H_6O_3).$$

With regard to the bacteria and the conditions under which they develop, further information will be given in dealing with the bacteria of milk. Milk-sugar was discovered in milk as early as the end of the seventeenth century. It is a crystalline body (four-sided prisms) of a white transparent colour, and contains, as has been indicated by the formula, one part of water of crystallisation. It is difficultly soluble in water or alcohol, and hence possesses only a slightly sweet taste. The brown coloration which milk is seen to assume, when heated to a comparatively high temperature, is due to the decomposition of this body. When heated between 100° and 131° C. (212° Fahr. and 268° Fahr.) it shows signs of decomposition and becomes brown. At 131° C. (268° Fahr.) it loses its water of crystallisation and undergoes further decomposition, which gives rise to the formation of galactine and perhaps also grape-sugar. With an increase of temperature the brown coloration deepens, and at 175° C. (347° Fahr.) the formation of a dark brown substance called *lactocaramel* takes place. In a watery solution of sugar, such as in milk, this decomposition takes place when the temperature rises above 70° C. (158° Fahr.).

The percentage of milk-sugar in milk may be said to vary between 3 and 6 per cent, amounting, on an average, to 4·68 per cent. Another carbohydrate is supposed also to be present in milk, but regarding it we know very little.

Ratio of the different Constituents of the Milk.—With regard to the variation in the different constituents, it may be added that the relation of the casein to the albumin is more or less variable; the fat is rarely less than the casein and albumin. The average proportion is 1·2 lb. of fat to 1 lb. of casein and albumin, or 1·5 lb. of fat to 1 lb. of casein. A milk containing less than 1·3 lb. of fat for each 1 lb. of casein has in all probability been skimmed. Milk with 3 per cent of fat will usually contain less than 12 per cent of total solids.

Mineral Constituents.—The mineral constituents of milk, despite their comparatively small proportion, have an important influence on its nature and properties. Their amount may be said to vary less than that of any of the other ingredients, and may be stated, on an average, at ·75 per cent, the limits of variation being from ·5 to ·9 per cent. They consist of *potash, soda, lime, magnesia,* and *iron,* in combination with *phosphoric, sulphuric, hydrochloric,* and *carbonic* acids. The three largest ingredients are potash, lime, and phosphoric acid, which are each of them present to the extent of from 20 to 26 per cent of the total ash. The following analysis (Schrodt) represents the average composition of the ash of milk:—

Potash	25·42
Soda	10·94
Lime	21·45
Magnesia	2·54
Ferric oxide	·11
Sulphuric anhydride	4·11
Phosphoric anhydride	24·11
Chlorine	14·60
	103·28
Deduct oxygen replaced by chlorine	3·28
	100·00

With regard to the condition in which the mineral matter is present in milk, it may be of interest to point out that the sulphuric anhydride present in the ash is derived from the sulphur, which constitutes a part of the caseous matter and albumin, while a portion of the phosphoric acid is formed by the oxidation of the phosphorus present in the casein to the extent of ·8 per cent. Again, a portion of the lime is, as we have already pointed out, in a state of combination with the casein; while lastly, a portion of the potash and magnesia is in combination with *citric acid* or other organic acids. With regard to the presence of this last-named acid in milk, it may be mentioned that Henkel has found it to amount to ·1 per cent, while according to the same authority the entire quantity of organic acids calculated as citric acid amounts to ·24 per cent. A portion of the magnesia, lime, and phosphoric acid are in a state of suspension in the milk.

The influence of the mineral matter in causing the so-called *amphoteric reaction* of milk will be referred to later on.

In addition to the above-mentioned substances, minute quantities of the following bodies have been found in milk, viz. *urea, lecithin, hypoxanthin, kreatin, chlorestin, leucin,* and *tyrosin;* no doubt formed by a partial decomposition of the albumin in the process of digestion in the animal body. The

presence of these substances, however, is of little importance.

Gases in Milk.—Lastly, we have *oxygen, nitrogen,* and *carbonic anhydride* in varying amounts. Of dissolved nitrogen the amount may be stated at ·75 per cent; dissolved oxygen ·1 per cent; free carbonic anhydride 7·5 per cent, and in a combined form from ·01 to ·2 per cent (Kirchner).

The constituents of milk are thus partly in a state of *emulsion* (the fat), partly in a state of *suspension* (caseous matter, lime, magnesia, phosphoric acid), and partly in a state of *solution* (albumin, milk-sugar, and most of the mineral matter).

The Reaction of Milk.—The reaction exhibited towards different vegetable colouring substances by milk varies. Human milk is normally *alkaline* in its reaction, while that of carnivorous animals is generally *acid*. With regard to cows' milk, however, the very curious fact has been observed that it exhibits the so-called amphoteric reaction; that is, both an acid and alkaline reaction. This is due to the presence in it of acid and neutral phosphates and carbonates of the alkalies.

Sheep's, Goats', and Mares' Milk.—With regard to sheep's, goats', and mares' milk, it may be mentioned that the first named possesses a yellowish-white colour and is characterised by a high percentage of total solids, which consist chiefly of fat and caseous matter; that of the second is almost pure white, and possesses a curious smell and flavour, being also slightly richer in fat and albumin than cows' milk; while that of the last named is bluish in colour, and possesses an aromatic sweetish taste.

The following table gives the average composition of the milk of the above-mentioned animals:—

	Sheep.	Goat.	Mare.
Water	83·0	85·5	90·7
Fat	5·3	4·8	1·2
Casein	4·6	3·8	2·0
Albumin	1·7	1·2	
Milk-sugar	4·6	4·0	5·7
Mineral Matter	·8	·7	·4
	100·00	100·00	100·00
Total Solids	17·0°/₀	14·5°/₀	9·3°/₀

Specific Gravity of Milk.—The specific gravity of cows' milk may be said to vary from 1·028 to 1·035 at 15° C. (water being taken as 1). This is the specific gravity of mixed milk. Occasionally, in the case of the milk of single cows, it may sink as low as 1·0263 or rise as high as 1·0380. As a rule, however, most milk will have a specific gravity varying between 1·030 and 1·033.

Colostrum or First Milkings.—As a general rule, the milk which the cow gives immediately after calving differs very materially in its nature from ordinary milk. So widely different is it in composition that it is known by a different name, viz. *colostrum*. This colostrum, or "beastings" as it is also called, is secreted, not merely after calving, but also

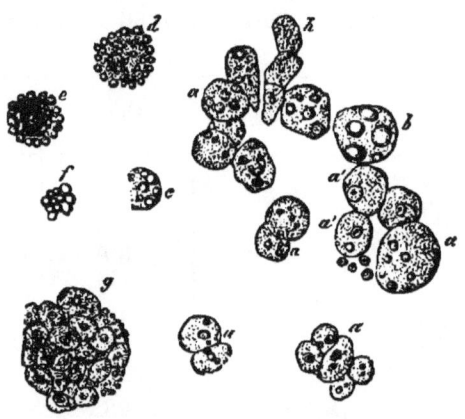

FIG. 9.—COLOSTRUM BODIES (× 300). *a'*, Cells with nucleus; *a*, cells undergoing fatty degeneration; *b*, cells containing large drops of fat; *c*, cells with a partially destroyed cell membrane; *d, e*, and *f*, cells which have entirely lost the cell membrane; *g*, cell masses from the milk canals.

for a short time before it; and not only does it differ from normal milk by the very much larger percentage of total solids which it contains—more than twice as much—but by the presence in it of bodies not found in normal milk. What is especially characteristic of colostrum is the presence of certain grape-shaped bodies, the so-called *colostrum corpuscles* (*corps granuleux*) (see Fig. 9). These corpuscles, which are four or five times the size of the ordinary milk globules, resemble in appearance white blood corpuscles, from which they have been considered by some to be derived. They disappear in milk, as a rule, in from three to fourteen days after calving, but sometimes not till the lapse of a much longer period. The most striking difference, between the composition of colostrum and normal milk, is to be found in the very much larger percentage of nitrogenous matter the former contains. Indeed, the richer quality of colostrum is due almost entirely to this fact, since the other constituents are not in excess of what they are in normal milk. The following is the average composition of the colostrum of 22 cows (Eugling) :—

Water	71·69
Total Solids	28·31
Fat	3·37
Casein	4·83
Albumin	15·85
Sugar	2·48
Ash	1·78
	100·00

From the above analysis, it will be seen that the increase in nitrogenous matter is due to the albumin, the casein being pretty much the same as in normal milk. Another point in which colostrum differs from normal milk, is the fact that it does not contain milk-sugar, or, if it does, only in mere traces. The milk-sugar is replaced by other kinds of sugar, viz.

grape-sugar or *lactose*. Colostrum further differs from normal milk in containing more ash, and in the composition of this ash. Nearly one-half of it (41·43 per cent) consists of phosphoric acid, as against about 27 per cent in the ash of normal milk. Globulin has been found in colostrum to the extent of 8·3 per cent (Emmerling), and such bodies as lecithin, cholesterin, urea, and nuclein have also been found in traces. Its composition, however, is a very variable one. It has been found immediately after birth to contain from 24 to 32 per cent of total solids. In three or four days, however, it becomes very similar to ordinary milk. Occasionally, even after the lapse of a fortnight, its peculiar properties do not entirely vanish. In appearance it is a turbid liquid, of a viscous or slimy nature, possessing a yellowish colour, a strong and peculiar smell, somewhat saltish in flavour, and, as a rule, a somewhat weak acid reaction. When boiled it coagulates, on account of the high percentage of albumin it contains. Rennet does not coagulate it, or, if it does, only very imperfectly. The fat of colostrum has a higher melting-point than the fat of normal milk, and is also distinguished from it by its peculiar smell and flavour. The large percentage of albumin it contains makes it a very nutritive and easily digested food for the young calf. Indeed, it seems to exert a special action on the alimentary canal. There are no grounds whatever for the belief, which has sometimes been popular, that colostrum is not suitable as a food for the young calf. The fact that, when allowed to stand, a cream layer separates out, which amounts to about 50 per cent of the whole, has given rise also to an erroneous but popular idea that it is very rich in fat. This has led to its use for churning. It need scarcely be pointed out that such a use of colostrum is to be condemned. As we have already pointed out, it is not so rich in fat as normal milk. It is, furthermore, highly unsuited for either butter or cheese making. For these reasons, the milk from cows which have recently calved should not be used, for from four to seven days, for churning, or for cheese-making, for double that period. Indeed, it is safer to keep the milk apart for even a longer time.

CHAPTER IV

Causes and Conditions Influencing the Quality and Quantity of Milk

THE causes which influence the *quality* and *quantity* of milk have been much studied, and although many points still remain obscure, a good deal of information has been collected on the subject. In the first place we shall discuss the causes which influence the quantity of milk secreted.

Causes Influencing the Quantity of Milk Secreted.—The amount of milk yielded by a cow depends on the activity of the milk-glands in the udder; and this, in its turn, is influenced by a number of conditions. For one thing, it seems to be dependent on the amount of milk the udder contains. Thus, so long as the udder is comparatively empty, and other circumstances are favourable, the secretion of milk by the milk-glands seems to take place unhindered. But whenever the udder becomes full of milk, the rate at which the secretion goes on is diminished. It still goes on, however, as is evidenced by the distension of the udder which takes place under such circumstances; but it is no longer at the normal rate. The pressure to which the udder is subjected seems to affect the nature of the secretion, the quality of the milk secreted being different.

Now, what has been above stated seems to throw much light on the interesting and well-known fact that, if a cow which has been in the habit of being milked twice a day is milked oftener, the result is not only that more milk, but milk of richer quality, is obtained than was formerly the case. But when the secretion of milk is forced to go on under pressure, not merely is the quantity of the milk lessened, but, owing to this pressure, more frictional resistance is offered to the passage of the fat globules, and the rest of the solids of the milk, through the secreting vessels and ducts in the udder. It is for this reason that milk secreted under pressure is poorer in quality. The above facts also explain another interesting phenomenon familiar to all interested in dairying, viz. that the milk which is first drawn when the cow is being milked is invariably poorer in quality than that last drawn—the so-called strippings, which, it is well known, are always very rich in quality. Indeed, the following figures, which contain the results of an experiment carried out by Boussingault, will illustrate that there is a steady increase in the total solids during the whole period of milking. In this experiment the milk drawn from the cow's udder was divided into six consecutive portions and analysed. The following were the results found :—

	(1)	(2)	(3)	(4)	(5)	(6)
Solids	10·47	10·75	10·85	11·23	11·63	12·67
Fat	1·70	1·76	2·10	2·54	3·14	4·08
Solids not Fat	8·77	8·99	8·75	8·69	8·49	8·59

From the above figures, it will be seen that the increase is limited to the fat, the solids not fat remaining practically uniform. A view which was formerly held to explain this was that the milk underwent a sort of creaming process in the cow's udder; the last drawn being the surface layer of the milk, and therefore richer in fat than that first drawn. In view, however, of our present knowledge of the structure of the udder, it is not likely that any such creaming process could take place. The explanation first given commends itself as a far more probable one.

Difference in Morning and Evening Milk.— If, in the case of a cow milked twice a day, the intervening period between each milking is the same, and if other conditions are similar, it may be said that there will be approximately no difference between the morning and evening milk, both as regards quantity and quality. If, on the other hand, the periods intervening between the times of milking be unequal, it will be found that the milk obtained after the longer interval is greater in quantity, but poorer in quality, than that obtained after the shorter interval. We may add that experiments have shown that when a cow is milked three or four times a day an increase in its amount to the extent of 20 per cent, and an increase in its fat to the extent of 25 per cent, may be obtained over that gained when it is only twice milked.

Another condition which influences the activity of the milk-glands in the udder is the *age* of the cow. The maximum development in respect of milk production in a cow is generally to be found between the fourth and fifth calf it gives birth to. The influence of food is also important, but we shall treat it under the conditions which influence the quality of milk.

Conditions Influencing the Quality of Milk.— From what we have pointed out it will be seen that the condition which influences not merely the quantity, but also the quality, of milk to the greatest extent, is the nature of the milk-glands—that is to say, the *individuality* of the cow. Next to the individuality of the cow may be said to be the *breed*.

Breed.—In choosing milking cows it is important to choose good milking breeds. Although it has undoubtedly been found that those breeds which give a large amount of milk do not, as a general rule, give as rich a milk as those giving less, it does not necessarily follow, as is too commonly believed, that cows yielding large quantities of milk yield poor milk; for it is quite possible to have both a large yield of milk, and one of good quality.

All farmers are aware that, so far as the quality of milk is concerned, no cows yield so rich a milk as the Jersey and Guernsey cows. Analyses of the milk of cows belonging to these breeds are on record which show an extraordinarily large percentage of total solids. As an example of such milk, reference may be made to a sample analysed by the author, in which the total solids amounted to no less than 18·39 per cent. The milking records made in connection with the London Dairy Show (1880-89) also illustrate this. Thus :—

	Total Solids.	Fat.
Shorthorns	12·87	3·73
Jerseys	14·65	5·02
Guernseys	14·23	4·90
Ayrshires	13·43	4·15

But a more convincing evidence of this difference is afforded by comparing the average of the results of a large number of analyses of the milk of different breeds; and interesting statistics, recently published in America, may be cited in this connection.[1] The results have been drawn from hundreds of analyses of samples of milk obtained from six different breeds. They are as follows :—

[1] *American Experiment Station Record*, vol. v. No. 10, p. 945.

	No. of Analyses	Total Solids.	Solids not Fat.	Fat.	Casein.	Milk-Sugar.	Ash.	Nitrogen.	Water.	Daily Milk Yield.
		per cent.	per cent.	per cent.	per cent.	per cent.	per cent.	per cent.	per cent.	lbs.
Jersey	238	15·40	9·80	5·61	3·91	5·15	0·743	0·618	84·60	14·07
Guernsey	112	14·60	9·47	5·12	3·61	5·11	0·753	0·570	85·39	16·00
Devon	72	13·77	9·60	4·15	3·76	5·07	0·760	0·595	86·26	12·65
Ayrshire	252	13·06	9·35	3·57	3·43	5·33	0·698	0·543	86·95	18·40
American Holderness	124	12·63	9·08	3·55	3·39	5·01	0·698	0·535	87·37	13·40
Holstein Friesian	132	12·39	9·07	3·46	3·39	4·84	0·735	0·540	87·62	22·65

From the above table it will be seen that the total solids ranged from 15·40 per cent in the case of the milk from the Jersey, to 12·39 per cent in the case of the Holstein—a difference of 3 per cent. As regards the fat, it will be seen that there is a difference of 2·15 per cent, the difference in total solids only amounting to 3 per cent, which leaves less than 1 per cent for the variation in all the other constituents; or, to put it in another way, the percentage of fat in the total solids ranged from 28 to 36·4 per cent, while the percentage of "solids not fat" ranged from 83·2 per cent to 63·6 per cent. The variation in the solids not fat only amounted to ·71 per cent, viz. from 9·80 to 9·07; and this variation was equal in amount in the case of the casein and milk-sugar, which varied between 3·39 per cent and 3·91 per cent, and 4·84 to 5·33 per cent respectively. The constituent which was most constant was the ash, which only varied between ·76 and ·69 per cent.

Period of Lactation.—Another influence affecting the quality of milk is the period of lactation—that is, the length of time after calving. And here it may be pointed out that the duration of the lactation period varies very considerably in the case of different cows. In very rare cases, the secretion of milk may actually continue without intermission from one calving period to another; while, on the other hand, it may occasionally cease a comparatively short time after calving. On an average it may be said to last some three hundred days. An American expert who has studied the question at considerable length comes to the conclusion that [1] "on an average cows give the thinnest milk just after calving; it becomes slightly richer during the next two weeks, and then it holds almost uniform in quality for four or five months, after which it gradually increases in richness as the cow comes near to calving again, and by the ninth month from last calving is only about one-seventh richer than it was during the earlier months." The difference in the quality of the milk due to this cause is manifested almost entirely in the fat—"the solids not fat" remaining practically the same. E. H. Dean, another American expert, has found that, dividing the lactation period into three parts of ninety-one days each, there was an increase in fat of only 17 per cent in the second period, and 46 per cent in the third period, over that of the first period.

Food.—The nature of the food may be mentioned as a condition influencing the quality of milk. This influence has been in the past both over and under-rated. The old belief was that food influenced, in a very direct manner, the composition of milk. The modern tendency, on the other hand, is to rather underestimate the effect that feeding has on milk

[1] W. W. Cook in *Agricultural Science*, vol. 7 (1893), pp. 253, 265.

secretion. While it is true that experiments have shown that change of diet seems to have little immediate effect on the milk, it is a mistake to imagine that milk is independent of feeding. There is no doubt that such conditions as individuality and breed, under ordinary conditions of feeding, count for more than food; still the importance of having a properly adjusted diet for milk cows is very great, and unless cows are properly fed they cannot be expected to yield their maximum quantity of milk, since under an ill-assorted diet the milk-yielding capacity is bound to suffer just as much as any other function of the body. Unfortunately, the experiments carried out on the influence of food have been comparatively few in number, and the results obtained very contradictory. There are, however, certain conclusions which we are warranted in making on the influence of food.

With regard to the three food nutrients,—the albuminoids, fat, and non-nitrogenous bodies,—the first named have by far the most important influence on milk. Whatever theory may be adopted as to the formation of milk in the cow's udder, there can be no doubt that the nitrogenous bodies in the milk are derived from the nitrogenous bodies of the cell substance. The food nutrient from which this cell substance is built up is bound, therefore, to have a very important influence on the composition of milk. The more abundantly the cell substance is built up, the more abundant will the secretion of milk be. An insufficient proportion of albuminoids, therefore, in the food will have the effect of diminishing the yield of milk. On the other hand, neither the fat nor the non-nitrogenous bodies in food are of such importance in influencing the composition of milk. So far as we can see, the fat has no effect on the production of fat in the milk. It may, however,—no doubt does, —exercise an important function in protecting the albuminoids from oxidation, and thus leaving them free to carry out what is their proper work, viz. to build up the cell substance. The influence which insufficient feeding is bound to exercise on milk formation, is towards impoverishing the milk yield in quality as well as in quantity.

Influence of Excitement.—Other causes influence the composition of milk, such as the condition of the cow when milked. At a meeting of the Society of Public Analysts in London, two years ago, a number of analyses of abnormal milk samples were read and discussed. Among the cases cited was one of a cow which had been milked at a fair when in an excited condition. The milk, on analysis, was found to contain only 10·85 per cent of total solids, and only 1·85 per cent of fat. That this was due to the excited condition of the cow was shown by the fact

that, when milked next day, when quiet and in normal condition, the milk showed 12·75 per cent of total solids and 3·64 per cent of fat. That the tendency of excitement may not always be in the direction of giving milk of an abnormally poor quality is illustrated by two other samples of milk obtained from cows at a fair, which showed, on analysis, respectively 19·5 per cent of total solids and 11·06 per cent of fat, and 16 per cent of total solids and 7·37 per cent of fat.

Illness.—Another cause which affects the quality of milk is illness. To what extent milk may be affected by this cause is illustrated by a case quoted by a German expert in 1893. Three samples of milk, obtained from a cow suffering from illness, were found to have as poor a composition as the following:—The solids ranged from 8·31 per cent to 9·16 per cent, the fat from ·25 per cent to 1·55 per cent, and the milk-sugar from 3·82 to 4·10 per cent.

Among other causes which influence the general quality of the milk may be mentioned exertion of a severe kind and the time of day. The season of the year has also an important influence, milk being generally poorer in spring than in autumn.

CHAPTER V

The Changes which Milk undergoes

THE number of changes which are constantly going on in milk are borne witness to by the rapidity with which that valuable food sours, coagulates, changes its colour, etc., when kept for any time. These changes are, many of them, of a very complicated nature, and as yet little understood. Although not always apparent, they start from the moment the milk leaves the udder. Most of them are connected with the action of the ubiquitous micro-organic life, which is being shown more and more every year to be of the highest importance in many operations. In milk, of all fluids, they find their most congenial home and an abundance of food, and they are not long in establishing themselves in such congenial surroundings. Some limit their attention to one milk constituent, and some to another, so that in a very short time a slow transformation of these different constituents is begun, which eventually leads to the changes in the properties of milk so familiar to all. In this Chapter we shall only casually refer to the micro-organisms which effect these changes, and shall leave a more full treatment of this important subject to a following Chapter.

Creaming of Milk.—Milk, as it comes from the udder, may be described as of practically uniform composition. If, however, we let it stand at rest for some time, we find that the uniformity of its composition is disturbed by an accumulation, which takes place more or less quickly on its surface, of its minute fatty globules. Concurrently with this separation of the fat, a change in the colour of the main body of the milk will be observed. By the removal of the fat the opacity of the milk is diminished, and it is rendered more transparent. This has the effect of imparting the bluish tinge, so characteristic of skim milk. The surface layer of milk, which is thus enriched in fat, is known as *cream*. Now there is a popular belief to the effect that the richness of milk in fat is indicated by the depth of this cream layer; and while, no doubt, this is true, within certain limits, it is, as we shall immediately proceed to show, not necessarily so. The tendency which the milk globules possess to rise to the top exists to such an extent that if a quantity of milk six inches deep be allowed to stand undisturbed, at a temperature of 15° C. (60° Fahr.), for twenty-four hours, from 80 to 85 per cent of the fat will be found in the surface cream layer. The rate at which the fat globules find their way to the surface is dependent on their size. The larger globules rise first. The very small globules never rise at all, as the amount of nitrogenous matter which adheres to them is too great for them to assert their lighter specific gravity. This accounts for the fact that all the fat is not found in the cream layer, even after milk has stood for a long time. No doubt, with the help of centrifugal force, more complete separation may be effected; but even under such circumstances complete separation is not obtained, although as much as from 90 to 96 per cent of the fat may be separated in this manner.

Souring.—After a time milk spontaneously coagulates and develops a sour taste. Before this takes place, however, a careful inspection of the milk would show that it had undergone very considerable changes. From the moment of its leaving the udder it is taken possession of by a class of bacteria which are known as lactic bacteria, and of which no less than one hundred different kinds have been identified already. These bacteria produce lactic acid in milk. This they do by decomposing the milk-sugar, which is present to the extent of about $4\frac{1}{2}$ per cent, and which is a very easily decomposable substance. Now these bacteria, like all other bacteria, are very much influenced in the rate of their development by temperature. This, indeed, is the reason why the influence of heat is so enormous in regulating the changes which take place in milk. Rapid development of these classes of bacteria may be said only to take place at temperatures above 15° C. (60° Fahr.). It is for this reason that milk ought to be cooled down, immediately after milking, below this temperature, if it is desired to keep it for any time. Similarly, the temperature ought not subsequently to be allowed to rise above this point. But, just as a low temperature is unfavourable for the development of bacteria, so also is a high temperature. Thus, heating the milk to 50° C. (122° Fahr.) checks this kind of fermentation. The effect of the formation of this lactic acid, when it reaches certain proportions in the milk, is to cause the coagulation of another important constituent of milk, viz. the caseous matter.

And here reference may be made to the widely spread belief that electricity in the air has the effect of souring milk. There can be no doubt that in "thundery" weather milk or cream sours more quickly than under ordinary conditions; but this cannot be traced to the effect of electricity. In fact, experiments have shown that electricity, instead of accelerating the souring of milk, actually retards it. The true explanation is probably to be found in the fact that the temperature of the air is generally increased before thunder, and that this promotes a more active development of bacteria in the milk. That it is not due to electricity is proved by the fact that *sterilised* milk will *not* sour, no matter how violent the thunderstorm is.

Amphoteric Reaction.—And here it may be well to say a word or two on rather a curious property of milk. This is the amphoteric reaction it possesses— a property which we have already referred to. By warming the milk, it is found that the alkaline reaction becomes more pronounced. Warming, however, has no influence on the acid reaction. By the gradual formation of free lactic acid in the process of fermentation, milk loses its amphoteric reaction, the alkaline reaction disappears, and the acid reaction alone remains and gradually increases in strength. This takes place, after a time, to such an extent that, although the milk remains liquid at ordinary temperatures, a slight increase of temperature, or the addition of carbonic acid, causes immediate coagulation. Finally, the casein, even at the ordinary temperature, becomes coagulated.

Coagulation of Milk.—The coagulation of the casein is thus due to the development of acidity in the milk. The higher the milk is heated—up to a certain point—the more quickly will coagulation take place. The coagulation of the casein may be effected by the addition of various precipitating reagents, such as dilute acids, salts, alcohol, etc., and by *rennet*. It is worthy of note that the coagulation, which is formed by spontaneous souring of milk, and that formed by different coagulating reagents, differ in their nature—a point which will be discussed in a subsequent Chapter.

Effect of Heat.—The extent to which coagulation of the casein in milk takes place and the relative effect of different coagulating reagents is influenced by temperature and certain other conditions. Thus the higher the temperature (up to boiling-point), the less the quantity of acids or other precipitating reagents required to effect coagulation. Again, coagulation seems to be greater the sooner the milk is treated after milking. The amount of precipitant to be used, therefore, will largely depend on the temperature. The action of heat on the properties of milk is altogether of a most interesting nature. When milk is heated to boiling or to higher temperatures, it develops a peculiar flavour—a flavour especially unpleasant to many people. Indeed, heating milk to even a considerably lower temperature than boiling temperature imparts this "cooked" flavour to milk. Thus if milk be heated to 158° Fahr. for fifteen or twenty minutes it acquires it. If, however, the milk be cooled again it loses it. It is possible that both the smell and flavour of milk which has been strongly heated is due to the presence of small quantities of sulphuretted hydrogen formed by the decomposition of such albuminoids as contain sulphur. The colour of milk is also slightly changed under the action of heat; and this is due to the decomposition of the milk-sugar and the formation of small quantities of yellow and brown substances of the nature of *lactocaramel*. The result of heating milk is to change the fine state of division of the fatty globules, many of which run together and form larger globules. Again, on heating milk, even to the comparatively low temperature of 50° C. (122° Fahr.), a skin is formed on the surface, caused by the coagulation of the albumin.

Milk Faults.—Milk has been found in the past to be liable to strange diseases, or, as they are technically known, "faults." Thus, for example, it has been found to develop strange coloured patches — blue, yellow, or green — or to have its whole colour changed. These milk faults, as a rule, only become apparent in the milk some time after it has been drawn from the cow's udder. The practical importance of the subject consists in the fact that a serious disturbance in the quality of the milk products is the result. These changes, as we shall see in the next Chapter, are generally caused by bacteria. Milk, again, is sometimes obtained in which the cream rises very slowly. The milk of cows which have been long milked often shows this property. It arises from the fact that the original condition of the nitrogenous matter in the milk becomes changed in an extraordinary manner, and a large proportion of the fatty globules become free. It has also been noticed that such milk contains less calcium phosphate than normal milk.

Means of preventing Changes in Milk.—It may be asked, How may these changes be prevented or retarded? Now, in doing so, the great agent is heat. Cleanliness is not a less valuable instrument : cleanliness in every way—on the hands of the milker, on the teats of the cow, in the milk-pails and other receptacles used for holding the milk, in the byre, etc. Immediately after milking, the milk should be cooled down : the lower the temperature the better. On the other hand, it may be sterilised by heating ; and, in order to avoid imparting the disagreeable boiled flavour to the milk, this may be effected by heating it to 70° to 80° C. (158° to 177° Fahr.). The addition of chemicals, so-called "preservatives," cannot be too strongly condemned. Even such comparatively harmless preservatives as bicarbonate of soda, boracic acid, salicylic acid, and peroxide of hydrogen, ought not to be used.

The methods of sterilising milk and the importance of cleanliness will be further referred to at the conclusion of next Chapter on the bacteria of milk, so that all further consideration of the subject may be postponed till then.

Preserved Milk.—The sterilisation of milk by submitting it to a high temperature, while not always necessarily effective, is accompanied by the objection that it imparts to the milk a flavour peculiarly disagreeable to many persons. On the other hand, sterilisation by "intermittent sterilisation" is an extremely inconvenient method, and is utterly unsuited for practical work. Many attempts have been consequently made in the past to convert milk into some form in which it will keep without undergoing decomposition. These attempts have all taken the form of preparing milk in a condensed form. At first preserved milk was sought to be prepared by evaporating milk to dryness. The solids thus left behind, after being mixed with a small quantity of bicarbonate of soda, were then pressed into the form of cakes. This method, however, did not meet with success. In the first place, the cakes were found not to keep well, since the fat soon acquired a rancid flavour. It was also found that such cakes did not dissolve properly in water, the drying process having destroyed the colloidal condition of the caseous matter. It was subsequently discovered that, by simply condensing the milk to about one-half or one-third of its volume, a condensed milk which kept fairly well could be prepared, and that the addition of cane-sugar still further increased its keeping properties.

As at present manufactured, condensed milk is prepared both in a sweetened and unsweetened condition, the sweetened variety having about 12 per cent of cane-sugar added to it. In preparing the unsweetened variety the milk is generally first sterilised by heating and then condensed in vacuum to a half or a third of its bulk. The following analyses of both varieties will illustrate their composition (Hehner):—

	Water.	Fat.	Albuminoids.	Milk-Sugar.	Cane-Sugar.	Ash.
Sweetened variety—						
Norwegian	28·85	9·21	8·98	14·14	36·74	2·08
Nestlé's Swiss Milk	15·30	8·85	9·98	13·62	50·08	2·17
Unsweetened variety—						
American (mean of ten analyses)	45·59	15·67	17·81	15·40		2·53

CHAPTER VI

The Bacteria of Milk

ALTHOUGH the functions performed by micro-organisms in the dairy have only comparatively recently come to be recognised, it has long been realised, by the help of experience, that a very necessary condition of successful dairying is *cleanliness*. By working in a cleanly fashion in the dairy, contamination with bacteria is avoided to a large extent; for the only dangerous constituents of dirt are these minute living organisms—dead dust is a harmless enough article.

The changes which micro-organisms may effect in milk have been again and again exemplified by the mysterious development of strange milk "faults," already referred to in the preceding Chapter. Among these may be mentioned such as manifest themselves by the milk becoming ropy, stringy, or viscous; by its assuming a blue, red, yellow, etc., colour,—or, at least, the development of coloured patches in it,—a bitter taste, and by its becoming liable to fermentative curdling, etc. It was, till lately, customary to regard such faults in milk as due to the nature and condition of the food, the nature of the soil, and, above all, to illness or disease in the cow—to everything, in short, except the true cause, viz. dirt. We now know, however, that such faults in milk are due to dirty and careless handling, or, more exactly stated, to micro-organisms which gain access to the milk when it is dirtily treated. Indeed, the best proof of this is to be found in the fact that improved dairy practice has rendered such faults of extremely rare occurrence. In large and modern dairies they are practically unknown, and it is only in small dairies, where unfortunately extremely primitive methods too often obtain, that they still occur.

Occurrence of Bacteria.—Before more particularly dealing with these micro-organisms, let us see where they are to be found. To this a simple answer may be given. Modern investigations have shown that their presence is well-nigh universal. In the air, in the soil, and in water they are widely distributed. In the air they are to be counted by thousands in every cubic yard, while in like quantities of water and the soil millions of them exist. It is true that, under certain circumstances, air may be, if not absolutely, yet comparatively, free of them. For example, it would seem highly probable that in air above the sea, far from land, they do not occur at all; while air on the top of high mountains is correspondingly pure. In uninhabited spaces their number is enormously less than in inhabited spaces. In the air above the streets of Paris it has been calculated that they are present to the extent of about 4000 per cubic yard.

Their tendency is to obey the law of gravity, and to subside from the air. This, however, can only completely take place when the air is kept absolutely undisturbed—a condition which, we need scarcely say, hardly ever obtains. But, numerous as they are in outside air, their number in that of enclosed spaces, such as the dairy, and more especially the byre, is even greater. Thus, one investigator has found 120 bacteria and moulds in a quart of the air of a byre. In the litter, in the manure, and in the dirt on the floors of byres they are very abundant, and whenever this is stirred or disturbed in any way large numbers of them are sent into the air. It is for this reason that the hands of milkers, the teats of cows, and the milk vessels are liable to become so rapidly contaminated with bacterial life.

Contamination of Milk by Micro-Organisms.— The thousand-and-one ways in which milk may thus become contaminated makes it a very difficult task to obtain it from the cow absolutely *sterile*,—that is, devoid of bacterial life,—or even to believe that milk when it comes from the udder is devoid of it. In many cases, even where the most elaborate precautions have been taken by experimenters, the attempt to draw sterile milk from the cow's udder has failed. And yet the doctrine, that milk as originally formed in the cow's udder is sterile, is a true one; that is, except when the udder is in a diseased state. The failures above referred to are not to be wondered at when we remember the difficulties to be contended with. In the first place, we have the difficulty of keeping the hands absolutely free from bacterial contamination. Then the comparatively wide surface of the milk brought into contact with the germ-laden air in the process of milking renders the task still more difficult. And lastly, assuming that such difficulties are successfully overcome, we have the difficulty of keeping the teats absolutely free from micro-organic life. It has been found, for example, that bacteria present in a drop of milk at the opening of the teat may even work their way into the milk cistern, and, thanks to the high temperature which there prevails, may very rapidly develop. Their complete removal from the udder may, consequently, not be effected till milking has progressed for some time.

As illustrating this point, Lehmann has found that the first 10 oz. of milk drawn contained from 50,000 to 100,000 per cubic centimetre (broadly speaking, the $\frac{1}{28}$th of an ounce), the main portion of the milk about 5000, and that it was only the last 10 oz. drawn that were practically free from bacteria.

In view of what has been now stated, we may affirm that milk, as usually obtained, contains micro-organisms, and that these organisms will develop and increase in milk, unless it is specially submitted to the action of heat, or is treated with antiseptics.

Importance of a Knowledge of Bacteriology to the Farmer.—Now, while bacteria may be regarded, as far as milk itself is concerned, as undesirable, yet we must remember that the manufacture of dairy products, such as butter and cheese, are dependent on their action; so that a consideration of the subject is doubly interesting to all engaged in dairying, not merely on account of the harm they do in promoting undesirable changes in milk, but also for the indispensable rôle they perform in the manufacture of butter and cheese. We should try to know, therefore, as much about the nature of these minute organisms, and about the conditions which influence their development, as possible, so as to enable us to take precautions either in checking or killing them, or in regulating and fostering their development when required. The success of the dairy industry depends on properly controlling certain fermentative processes, such as that potent in cream-souring for butter manufacture, and that active in starting and developing certain ripening processes in cheese manufacture. We want, therefore, as much information as to how we can best foster these kinds of fermentation as can possibly be obtained. If the information we already possess regarding these minute workers in the dairy is meagre, we must recollect that it is always being added to, and that its value is great.

Milk as a Propagator of Disease. — There is another aspect of milk, considered from a bacteriological point of view, which merits attention, viz. the risk it runs of acting as the propagator of various kinds of disease. The peculiarly rich properties of milk as a food renders it admirably adapted for forming a nutritive medium for the development of various disease (the so-called pathogenic) germs, which are constantly to be found, along with other micro-organisms, in the air and elsewhere. There can be no doubt whatever that such diseases as typhus, diphtheria, cholera, and, above all, consumption, which are caused by germ life, are too often disseminated in this way. Indeed, it would seem highly probable that a very large quantity of milk used is infected with the tubercular bacillus (see note ^{a}, p. 113), and that in this way the seeds of that most insidious of all diseases, consumption, is implanted in the system, more especially in the case of children. According to certain authorities, about 5 per cent of the samples of town milk contain tubercular bacilli. As to the method in which the tubercular bacillus obtains access to the milk there can be no doubt. It does not come into the milk, as a rule, subsequent to milking, but is already in the milk when it leaves the udder. The enormous prevalence of tubercular disease among cows—it has been calculated that $2\frac{1}{2}$ per cent of all cows butchered are infected with tuberculosis—renders the danger of milk acting as a propagator of this disease very great. It must not be inferred, of course, that all milk containing tubercular bacilli is equally capable of producing the disease. The probabilities are that where it is present in a sample of milk, this sample, before it is consumed, becomes mixed with a large quantity of milk free from bacilli of this type; but the resulting dilution, as has been very properly pointed out by Freudenreich, while it diminishes the risk of infection, does not do away with it entirely. Typhus, again, has often been spread by milk. In fifty typhus epidemics which Hart investigated in England, no less than twenty-eight were found to be due to infected milk. Here, of course, the infection would be imparted to the milk after milking, in any of the many numerous ways in which milk becomes contaminated. With regard to cholera, it has also been proved again and again that milk has been the means of propagating this terrible disease. While lastly, with regard to scarlet fever and diphtheria, Hart has traced no less than fourteen epidemics of the former, and seven of the latter, to this source in England; and while we have not the same evidence with regard to other infectious diseases, there can be little doubt that many of them in the past have been spread in this manner. There would seem to be a certain class of bacteria that are quite harmless when taken into the system, under ordinary conditions; if, however, the vitality of the body is in an impaired state they become a source of disease. Such bacteria are probably common in milk, and are doubtless the cause of gastric and intestinal disturbances so common in young infants during the summer months (summer diarrhoea and cholera infantum). They are able to produce certain by-products which have a poisonous effect when taken into the susceptible digestive tract of the infant.[1] The importance, therefore, of a knowledge of the bacteriology of milk, in the interests of public health, will thus be seen to be enormous.[b]

Although the existence of the world of micro-organisms was drawn attention to as far back as 1675 by Leeuwenhoek, who discovered them first in the saliva of the mouth, still it is only within the last thirty or forty years that any strides in the science of bacteriology have been effected. Still more recent is our knowledge of their enormous distribution in air and water, and our conception

[1] See Bulletin 44, *Ag. Exper. Stat.*, University of Wisconsin.

of the important rôle they perform with regard to human life. In view of this, the progress already made is most encouraging, and is a happy augury for the future.

A word or two of a more or less general nature may be in the first place said with regard to bacteria.

Description of the Different Micro-Organisms.— These tiniest members of organic life, which represent its lowest form, and of which about a thousand different species have been already studied, were formerly regarded as belonging to the animal kingdom, and were called by Ehrenberg in 1828 *Infusoria* or infusion animals. The tendency of the most modern research, however, is to classify them rather along with the plant world. They consist of a single cell, and grow like the cells of plants, which in many other respects they resemble. Indeed they are related by many links to the Algae. They may be divided into different classes. This classification is generally based upon their shape. Under the microscope they are seen to be joined to one another in chains, bundles, or heaps, and occasionally as firm glutinous masses. Those possessing a round globular form are known as *cocci, micrococci, macrococci* (when they occur singly), *diplococci* (when they occur in pairs), and *streptococci* (when they are arranged in chains), and lastly *staphylococci* (when they appear in grape-like bunches). Sometimes they are united together by a jelly-like material; and in such species as thrive in moist or fluid surroundings, the outer layer of the cell wall is of a jelly-like nature, and is provided in some cases with minute whip-like filaments, the *cilia*, by means of which they are enabled to swim about, and in this respect resemble animal life. The way in which such bacteria move differs considerably, some having a circular spinning motion, while others have a spiral motion, and others, again, glide along. When the organism is exposed and dried, this outer jelly-like layer becomes a hard crust. Those of a rod-like shape are known as *bacilli* or *bacteria*, the former being longer than the latter. Those possessing a spiral or corkscrew shape are known as *spirilla*, or when in a shorter form as *comma bacilli*. The term bacteria, however, is generally used in a wider sense to denote the micro-organisms above enumerated. In addition to these there are two other classes of micro-organisms: viz. *yeasts*, which are comparatively large oval bodies; and the *moulds* so familiar to all as giving rise to hairy patches on various kinds of food, especially jam (see Fig. 10, p. 73).

Method of Development of Bacteria.—A point of much interest, and one also of great importance, is the method in which bacteria develop. This is done in a variety of ways. Some grow in length, some increase by splitting up into two parts, which is known as *fission* (Fig. 11, p. 74), and some by throwing off *spores* (Fig. 12, p. 74)—small round or egg-shape bodies which are formed inside the bacterium, which

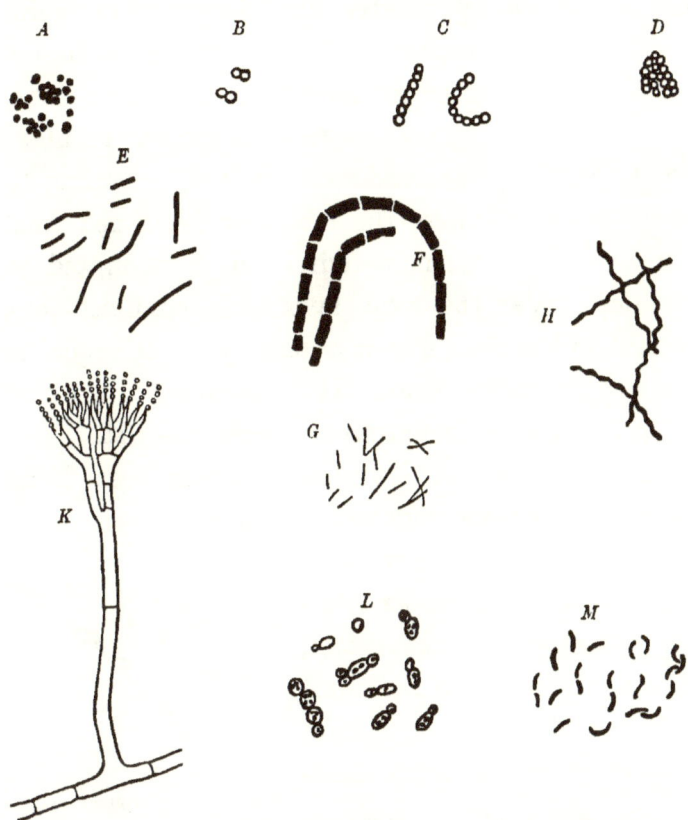

Fig. 10.—Different forms of Micro-Organisms. *A*, Micrococci (Frankland); *B*, diplococci (Freudenreich); *C*, streptococci (Freudenreich); *D*, staphylococci (Freudenreich); *E, F, G*, bacilli (Frankland); *H*, spirilla (Frankland); *K*, mould (*penicillium glaucum*, Freudenreich); *L*, yeast cells (Frankland); *M*, comma bacilli (Frankland).

in turn, under suitable conditions of heat and moisture, develop into full-grown bacteria. Now in this process of reproduction or development one or two points are worthy of note. For one thing, it may take place at an enormous rate. A bacterium may develop in twenty minutes' time the power of repro-

FIG. 11.—MICROCOCCI, SHOWING MODE OF MULTIPLICATION BY FISSION. (Frankland.)

ducing itself, so that under favourable circumstances, in a few hours' time, it may give rise to a progeny numbering millions. Indeed, according to Conn, the microbes descended from a single individual, if permitted to develop under the most favourable circumstances, would, in less than five days, occupy a space equal to that of the entire ocean. It is needless to say that such favourable circumstances never do exist.

Again, with regard to spores, it has been found that they possess very much greater powers of resistance than the developed bacterium. The greater vitality possessed by the spores, in comparison with the adult cells, is due to the harder nature of the covering membrane. The latter is, as a rule, comparatively easily killed by unfavourable circumstances, but the former in some cases possess extraordinary powers of resistance.

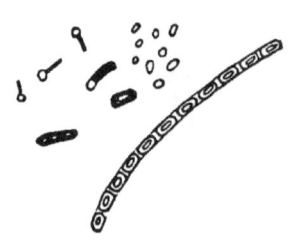

FIG. 12. — MULTIPLICATION BY SPORES OF BACILLI. (Frankland.)

Size of Bacteria.—The size of bacteria is so minute that they require a high power of the microscope to distinguish them. They are only visible to the naked eye when they occur in colonies in some fixed medium. Many of them are less than $\frac{1}{20,000}$th of an inch in length, and few of them greater than $\frac{1}{300}$th of an inch. Hundreds of millions of these organisms could be spread over a square inch in a single layer. A bacterium which causes lactic fermentation is only about $\frac{1}{25,000}$th of an inch broad and $\frac{1}{8,000}$th of an inch long. No less, therefore, than twenty-five thousand of these bacteria could be placed side by side without occupying more than an inch, and nine hundred billions of them would only weigh $\frac{1}{28}$th of an ounce.

Conditions influencing the Development of Micro-Organisms.—The conditions of life of micro-organisms is a point which it is very important we should consider, as our knowledge of this subject furnishes us with the means of regulating their development.

Bacteria may be divided into different classes, according to their products. Thus a large class give rise to pigments, another set up fermentation, and a third class give rise to putrefactive decomposition.

We can, however, divide them into two great classes, according as they live on dead matter or on living organisms. To the latter class belong the pathogenic—the disease-producing—germs. Among diseases that have been proved to be due to bacterial life may be mentioned tuberculosis (consumption), diphtheria, erysipelas, lockjaw, pneumonia, typhus, cholera, glanders, malaria, and leprosy. The accompanying illustration shows the form of some of these pathogenic germs (see Fig. 13, p. 77). Their pernicious action is due to poisons, the so-called ptomaines, which are produced within the body. The functions performed by the other class of micro-organisms are highly useful, and it is with them we are mostly concerned when dealing with the bacteria of milk. Among the most important conditions in regulating their development is *temperature*. For most the range of temperature is generally between 15° to 40° C. (59° to 104° Fahr.), about 32° C. (90° Fahr.) being the most favourable temperature. Comparatively low temperatures (50° to 60° C.; 122° to 140° Fahr.) are sufficient to kill most bacteria. Similarly, low temperatures exercise a fatal effect. On the other hand, the spores of many bacteria have been found to possess great powers of resistance. Some of them are able to resist the lowest possible obtainable temperatures. Thus Pictet found that the spores of certain bacteria were able to survive in frozen oxygen at a temperature of $-213°$ C. ($-353°$ Fahr.). Some can survive boiling temperature; while a few have been discovered that are able to resist even considerably higher temperatures, viz. dry heat of 150° C. (302° Fahr.). The important bearing which these facts have on the practice of dairying is obvious. They explain why heating milk to between 60° to 65° C. (140° to 150° Fahr.) for some time

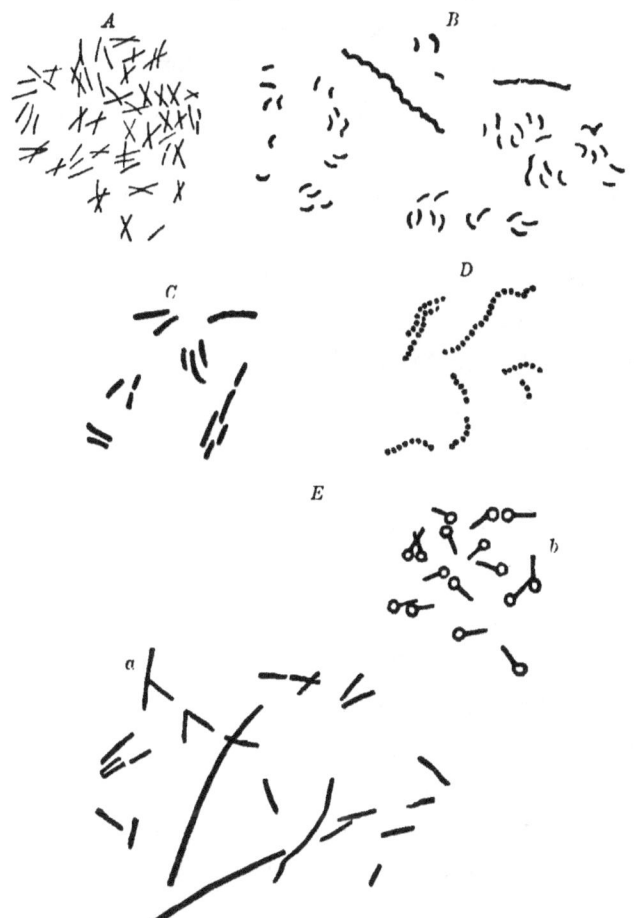

Fig. 13.—Pathogenic Germs. *A*, Bacillus tuberculosis (Koch); *B*, Spirilla of Asiatic cholera (Koch); *C*, Bacillus of typhoid fever (Migula); *D*, Streptococcus of erysipelas (Frankland); *E*, Bacilli of tetanus or lockjaw —*a*, individual bacilli; *b*, spores (Kitasato).

will do much to destroy bacterial life. And here it may be pointed out that there is a considerable difference between damp and dry heat, the former being very much more deadly than the latter in its effect. In order to make sure of effecting the complete destruction of the bacteria in any object, it is necessary to submit that object to a very high temperature for some time—a temperature of from, say, 160° to 163° C. (320° to 326° Fahr.). The fact that the spores possess greater powers of resistance than adult germs explains the efficiency of the method largely used for reducing bacterial life in such fluids as milk. This consists of heating the fluid from time to time at a comparatively low temperature, but sufficiently high to destroy adult germ life. When such a fluid is heated for the first time, many spores present in it escape destruction. As, however, the fluid is kept at a temperature favourable for germ development, the spores in due time develop into adult organisms, and are destroyed in the course of heating. Such a method of treatment, which is known as intermittent sterilisation, while it may be said to ensure complete destruction of micro-organic life, is yet too troublesome a method to have always recourse to. It is a fortunate circumstance that most pathogenic germs are comparatively easily killed. Generally speaking, it may be said that bacteria do not develop below 4° C. (39° Fahr.), and much above 50° C. (122° Fahr.). By cooling a liquid containing bacteria, therefore, to 4° C. (39° Fahr.), bacterial development is completely stopped; but, as a rule, by cooling to 7° C. (45° Fahr.) the same end will be effected.

Among other conditions which regulate the development of micro-organic life are the *reaction* and *nature* of the liquid or other medium in which they exist, the absence or presence of the *oxygen* of the air, and the absence or presence of *light*. With regard to the reaction of the liquid, it is a noteworthy fact that some bacteria only develop in the presence of a neutral or alkaline reaction, while others can develop in an acid reaction. Most bacteria require the former condition. An example of the latter kind are those producing lactic acid—the bacteria which induce the souring of milk. Acid reaction, we may add, is especially favourable for the development of moulds.

But one of the most important factors in regulating bacterial development is the nature of the nutritive medium in which they find themselves. Like higher organisms, they have their preference in the matter of food. They require oxygen, carbon, water, and mineral salts. Most of them also, however, require nitrogenous diet. It is hardly necessary to observe that, of all media, milk is the best, and hence it is that it becomes the happy hunting-ground of so many of them.

Their behaviour in the presence of oxygen serves to divide bacteria into two great classes. Some—

and to them the term *aerobies* has been applied—require air, and perish if they do not have access to it. Others, on the contrary, can develop without air. Recent researches of a most interesting kind have shown the important influence which sunlight exerts on bacteria. It has been found that, for most kinds of bacteria, sunlight is highly inimical. Indeed, putrefactive liquids may be actually rendered sterile by simply submitting them to the action of sunlight. A free supply of air at the same time helps sunlight to exert its full sterilising effect. Lastly, it may be added that moisture, as a rule, is a necessary condition, although its absence does not in all cases prove fatal. With regard to the development of yeasts and moulds, it may be explained that the former multiply by "budding" or "sprouting," while the latter develop by the formation of long threads (*hyphae*).

Having thus described the general nature and conditions of life of micro-organisms, we may next proceed to deal with those met with in milk. And before dealing with the different species, it will be well to say a word or two regarding their number.

Number of Bacteria in Milk.—We have already pointed out that in milk, even drawn under the most careful conditions, hundreds of thousands exist in every ounce. From 60 to 100,000 have been found in 1 cubic centimetre ($\frac{1}{28}$th of an ounce) of milk a few minutes after milking. Milk investigated by

Uhl in different cities showed from 3,338,775 to 55,365,800 per litre (1¾ pints). Their number rapidly increases on keeping, and it has been estimated that in milk which has stood six hours there may be from two to six millions, on an average, in one cubic centimetre. The temperature, of course, will affect their number. Conn has found, for example, that in a sample of milk kept for four days in a cold cupboard there were only 10,000 bacteria per cubic centimetre; while the same milk, taken from the cupboard and left in a warm room for six hours, contained 1,000,000. The following table shows the influence of temperature on their rate of increase:—

INCREASE IN NUMBER OF BACTERIA (CNOPF).

	At 34° C. (93° Fahr.)	At 12·5° C. (54·5° Fahr.)
In 1 hour	7½ fold	none
,, 2 hours	23 ,,	4 fold
,, 3 ,,	64 ,,	6 ,,
,, 4 ,,	215 ,,	8 ,,
,, 5 ,,	1830 ,,	26 ,,
,, 6 ,,	3800 ,,	435 ,,

As further illustrating the influence of temperature, it may be mentioned that a difference of 10° C. (18° Fahr., viz. from 59° to 77° Fahr.) made a difference in milk kept for 15 hours of 71,900,000 (viz. from 100,000 to 72,000,000) bacteria per cubic centimetre (Miquel). In milk kept for 24 hours at 25° C.

(77° Fahr.) no less than the gigantic number of 577,500,000 bacteria per cubic centimetre have been found (Freudenreich).[c]

Classification of Bacteria infesting Milk.— The bacteria infesting milk may be classified in various ways. The classification adopted by Grotenfelt [1] is the following :—

I. *Bacteria whose Action is Indifferent.*—In the first place, we have a large number of bacteria whose action is quite indifferent; that is to say, which exercise, so far as we know, neither a harmful nor a beneficial action.

II. *Bacteria whose Action is of a Useful Nature.*— Secondly, we have bacteria whose action in milk is of an indifferent nature, but which are active in milk-products—cheese, etc.

III. *Bacteria whose Action is of an Indirectly Injurious Nature.*—Thirdly, we have bacteria which exert an indirectly injurious action on milk. Such bacteria produce conditions which are favourable for the development of bacteria whose action is directly injurious. This class includes bacteria which cause an alkaline reaction in milk—a reaction favourable to a large class, indeed the majority, of ferments. Under this class also come certain bacteria which exert a directly injurious action. For instance, the

---

[1] See *Principles of Modern Dairying*, by Grotenfelt, translated by F. W. Woll (John Wiley and Sons), p. 94.

bacteria causing acidity in milk may be included under this class, since they are the necessary fore-runners of other bacteria which only thrive in strongly acid media, and which are therefore most frequently found in sour milk. As an example of this class may be mentioned the butyric bacilli, which impart to milk a strongly bitter flavour, but which are most familiar through their action in rancid butter.

IV. *Bacteria whose Action is of a Directly Injurious Nature.*—Fourthly, there are bacteria which exert a directly injurious action on milk, and which are therefore of greatest interest in this connection. This last class of bacteria may be divided into different sub-classes. Thus we have one kind which produces acidity, and of which the lactic bacteria and certain colour-producing bacteria, such as the *bacillus prodigiosus,* may be cited as examples. To this sub-class also belong bacteria producing butyric fermentation, which develop in the absence of air, and which produce a slight coagulation of the caseous matter of the milk, which becomes subsequently dissolved; as also bacteria producing volatile acids, under the action of which milk generally assumes a grayish colour and becomes rather viscous, while a lively generation of gas often takes place, consisting of carbonic acid, sulphuretted hydrogen, and, it may be, small quantities of alcohol and acetic acid. Another sub-class are those which produce no acidity in milk, but which effect the coagulation of the caseous matter, or which effect the fermentation of the caseous matter without coagulation. This class includes such bacteria as the *potato bacilli*, a group of bacilli to which Duclaux has given the name *tyrothrix*, and those bacteria which were formerly known as "putrefactive" bacteria, since they decompose albumin and produce an unpleasant smell.

Lastly, we have micro-organisms present in milk other than bacteria, viz. *yeasts* and *moulds*.

Before treating of the action of certain classes of bacteria which effect the common kinds of fermentation going on in milk, such as its souring, it will be convenient to refer to certain milk "faults" which have been shown to be generally caused by the action of bacterial life. Of these, the following are the most important :—

Blue Milk.—This fault consists in the development in the milk, after a lapse of from twenty-four to seventy-two hours, according to the temperature of the milk, of patches on its surface of a blue colour. It was the first kind of fermentation of milk to be studied, having been investigated by Foulkes in 1841. The formation of these blue patches only takes place—to any extent, at any rate—after the milk has assumed a slight, though distinct acid reaction. It completely ceases as soon as the milk has become coagulated, or, to speak more correctly, as soon as the caseous matter has become coagulated. It was on this account formerly thought that the cause of blueness in milk was connected with the casein, and that it was produced by some chemical ferment which acted upon the casein. We now know, however, that blueness in milk is due to the action of micro-organisms. The history of this discovery is an interesting one, and much research has been devoted to its elucidation. The organism chiefly implicated in the action has been called the *bacillus cyanogenus*.[d] It cannot produce a blue colour apart from milk, and seems to act on the casein and to have no effect on the milk-sugar. Indeed, it would seem that the explanation of blue milk is a double one, and that it is the joint work of the above-mentioned bacillus in conjunction with the lactic organism, since it has been found that if the *bacillus cyanogenus* be inoculated into sterilised milk, it does not produce a blue colour, this being only produced by the addition of a little acid,[e] or by inoculating the milk with lactic organisms. It thus seems to be an example of what is known as symbiosis, that is, the association of several species of micro-organisms, working together to produce some special effect. We have further examples of such "feeding" associations, in other departments of agriculture, such as in the case of the organisms which enable leguminous plants to utilise the free nitrogen of the air. Further evidence in support of this explanation is to be found in the method in which the blue patch first appears.

Where this organism comes from is not known, but there can be little doubt that its source is filth. Blue milk possesses an acid smell. It may be churned, but does not yield good butter; although it is a mistake to suppose, as was formerly done, that it possesses any poisonous properties. It may be added that it is killed by exposure to a temperature of 80° C. (176° Fahr.).

Red Milk.—This fault may be caused by red colouring matter in the food, or by the red colouring matter of the blood, which passes into the milk, or it may be caused by the action of micro-organisms. Where the food causes redness in milk it is generally due to the presence of madder in it. When the colour is due to blood, this may enter the milk from some small rupture in the udder. Lastly, where redness is due to the action of micro-organic life, it is generally caused by the development of a blood-red micro-organism, the *micrococcus prodigiosus*, which develops especially abundantly on potatoes. Among other micro-organisms which have been identified as causing it are the *bacillus lactis erythrogenes* and the *sarcina rosea*. The occurrence of such micro-organisms in milk is a very rare one. The bad influence they exert on the milk consists in the fact that they cause it to rapidly coagulate. In other cases they lead to the formation of *trimethylamin*, and produce a herring-like smell and flavour.

Yellow Milk.—This may be produced by the *bacillus synxanthus*— a micro-organism belonging to the casein group of ferments, to be subsequently considered — or other micro-organism. The organisms producing it do not act alike. Some produce a brilliant yellow colour, while others resemble rennet in their action, and first curdle the milk, the curd being then dissolved into a yellow or orange or amber-coloured liquid.

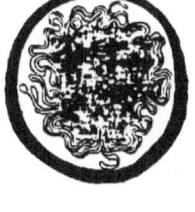

Fig. 14.—Colonies of Bacillus Violaceus. (After Grace C. Frankland.)

There are also certain kinds of organisms which turn the milk *green*, and some which turn it *violet*. As one of the causes of this last fault, the *bacillus violaceus* has been identified (Fig. 14).

Slimy or Ropy Milk. — Milk occasionally, instead of exhibiting its ordinary limpid nature, becomes thick and slimy. Such milk either does not cream at all, or else very imperfectly. It cannot be churned, and is not suited for drinking purposes. In some countries, however, — as, for example, in Norway, where it is really liked as an article of diet (Tattemyelk)—it is produced artificially by immersing the stem of the butter-wort (*Pinguicula vulgaris*) in the milk, or by feeding this plant to the cows. Slimy milk is used in the manufacture of one kind of cheese, viz. *Edam*. On the whole, however, it is regarded as an unmixed evil, and has caused great trouble in dairies. Various theories have been put forward with regard to the cause of slimy milk; but, like most other faults which we have mentioned, it has been finally traced to the action of micro-organisms. Pasteur was the first to discover that there was a special yeast which had the power of giving rise to a slimy fermentation of milk-sugar, and thus gave a proper direction to the investigation of the subject. Lister subsequently studied the question of slimy milk, and attributed it to the growth of bacteria. More modern researches on the subject seem to show that slimy or ropy milk is caused by a large variety of micro-organisms, among which may be mentioned the *bacillus mesentericus, leuconostoc mesenteroides, actinobacter du lait visqueux,* *actinobacter polymorphus, bacillus viscosus,*[1] *bacillus lactis pituitosi, potato bacilli, streptococcus Hollandicus, micrococcus Freudenreichii,* and *bacterium Hessii,*[2] as well as others.

It must not be imagined, however, that the nature of the fermentation is the same in all these cases. They produce a widely different effect on milk. Some induce only a slight amount of sliminess, while others have a very much more marked effect. The length of time they take to effect this slimy condition differs widely, and the products are also of a varied character. In some cases the substance produced seems to resemble cellulose in its nature, in others it seems to be of the albuminoid order. Mannit, carbonic acid, lactic acid, butyric acid, peptone, etc., are among the products. Altogether, no less than eighteen distinct organisms have been identified as associated with this slimy fermentation.

Bitter Milk.—A very common fault in milk is what is known as bitter milk. This may be, like many other faults, due to a number of different causes, but it is probably most commonly due to micro-organic life. In the first place, it may be caused by the nature of the food, which may contain some bitter

[1] In length from the $\frac{1}{20000}$th to $\frac{1}{14300}$th of an inch, and in breadth from the $\frac{1}{12700}$th to $\frac{1}{10500}$th of an inch.

[2] From the $\frac{1}{8800}$th to the $\frac{1}{5500}$th of an inch long, and $\frac{1}{30500}$th of an inch broad.

principle. This is probably the case only when the bitter taste is present in the milk immediately after milking; when it is developed only after standing for some time, it is due to some other cause. In such a case the cure is obviously to change the feeding. It may be yielded by cows which have been a long time in milk, and in such a case it is due to imperfect secretion of the milk in the milk-glands. Lastly, it may be due to inflammation of the udder.

The older theory that bitter milk is always accompanied by the production of butyric acid has not been shown to be absolutely correct; for, while it is at times associated with the production of butyric acid, it is also found to be produced by organisms which do not produce butyric acid. There are quite a number of species of bacteria which effect this result, and it would seem that in all cases the bitter flavour of the milk is independent of butyric acid. Some of these organisms resemble in their appearance the *proteus vulgaris*. To what, precisely, the bitter taste is due is not as yet known. One theory is (Hueppe) that it is connected with the decomposition of the albuminoid matter of milk, and with the production of peptones; and it is to these peptones and similar products that the bitter taste may be ascribed.[g] The fact that these bacteria chiefly belong to the casein group of ferments, which are characterised by their great powers of resistance, explains why it is that the bitter taste is most frequently found in boiled milk which has been allowed to stand for some time. In such cases, by boiling the milk, the bacteria effecting lactic fermentation are killed, and the spores of the bitter-milk bacteria only are allowed to remain. Such treatment is favourable to their development, since the presence of lactic acid in unboiled milk is inimical to their growth. Among the micro-organisms which have been identified as causing this fault may be mentioned Weigmann's *bitter-milk bacillus*, which possesses a length of $\frac{1}{16,600}$th to $\frac{1}{18,800}$th of an inch long, and from $\frac{1}{27,000}$th to $\frac{1}{22,700}$th of an inch broad; and Conn's *bitter-milk micrococcus*.

There are a number of other faults to which milk is liable, and which are due to micro-organic life, such as *premature curdling*, and the development of a *salt taste*, regarding which not much is as yet known.

Having discussed the different "faults" to which milk is liable, we may pass on to describe the cause of the more common changes which milk undergoes in keeping. The bacteria causing these changes may be most conveniently classified according to their products.[h] We have two principal groups, known as *Lactic* ferments and *Casein* ferments.

Lactic Fermentation of Milk.—The most common fermentative change in milk is its souring, which takes place on keeping milk for any length of time, and which results sooner or later in its so-called spontaneous coagulation. This souring is due to the formation of lactic acid, derived from the decomposition of the milk-sugar in the manner explained in a former Chapter. The production of lactic acid from milk-sugar may be effected by quite a large class of bacteria; and it has been found that while one ferment may be the common exciting cause in one district, another ferment may be the exciting cause in another district. In the case of some of these bacteria the lactic acid is merely incidentally formed along with other products. Such bacteria are those producing different colours in the milk, which we have already referred to. Another interesting fact with regard to this class of fermentation is that it is highly doubtful whether any two of the organisms effecting it act on the milk in exactly the same manner. An interesting observation made by Grotenfelt has shown that the lactic bacilli lose their power of producing lactic acid if they be cultivated for a time in a medium free from sugar. Again, the amount of lactic acid produced depends on the original condition of the milk, as well as the temperature. One of these ferments splits up the molecule of milk-sugar comparatively easily into four molecules of lactic acid, producing at the same time an extremely slight evolution of carbonic acid. Other lactic ferments produce small quantities of secondary by-products, especially alcohol. The most important practical application of lactic fermentation is in the souring of cream in the manufacture of butter; and as it is known that the various different lactic bacilli are not all equally adapted to act as ferments in effecting this change, attempts have been made to isolate and cultivate, in pure cultures, such as will produce the best butter with the finest aroma. A point of much interest in lactic fermentation is that it only continues for a certain time, this being the case with many other kinds of fermentation. When it has proceeded to a certain extent, the acid generated becomes fatal to the bacterial life causing it, and it ceases; consequently when much over one per cent of lactic acid is generated fermentation ceases. Curdling, it may be added, only takes place after the lactic acid has reached a certain amount.

A number of the bacteria causing lactic fermentation have been isolated and studied by different observers; among others by Hueppe, Grotenfelt, Marpmann, Krueger, and Weigmann. Thus Hueppe has discovered a bacillus about $\frac{1}{25,000}$th of an inch long and $\frac{1}{62,500}$th of an inch broad. It develops at temperatures between 10° to 45° C. (50° Fahr. and 113° Fahr.), and most rapidly at 35° C. (95° Fahr.). Other bacilli, micrococci, sphaerococci, streptococci, and bacteria have been studied by Hueppe and the above-mentioned investigators, which cause lactic fermentation. Lactic bacteria do not, as a rule, develop spores, and consequently have not very great resistant powers; a temperature of 70° C. (158° Fahr.) usually killing them. The chief difficulty of keeping milk is due to the presence of this class of bacteria; and hence it is that heating milk even to a comparatively low temperature has such an effect in preserving it.

Lactic ferments play a very important rôle in the dairy in the case of making butter from slightly soured cream. This is generally done by adding small quantities of sour milk to the cream, and then permitting it to stand for some time before churning. But here there is a danger, since it often occurs that the milk thus added to the cream contains bacteria which give rise to disagreeable odours and a bad taste. From a bacteriological point of view, therefore, it may be said that butter should only be made from sweet cream. Weigmann has attempted to prepare pure culture of cream-ripening ferment. In this attempt he has been successful, and he has succeeded in isolating a micrococcus which sets up normal lactic fermentation.

Certain of the lactic bacteria exert, as we have just indicated, a harmful action. This is seen in the manufacture of cheese. They split up the milk-sugar with such energy that gaseous products are evolved, and the cheese becomes perforated with holes. To one such class of bacteria Freudenreich has applied the name *Schaffer bacillus*. It seems to be closely related to a bacterium common in the intestine —*bacterium coli commune*. It is hence of the greatest importance that milk should not become contaminated, in the process of milking, with cow-dung.

Butyric Fermentation.—Occasionally it happens that milk or cream coagulates without any previous lactic fermentation. This is seen in the coagulation of milk which has been boiled, and the reaction of which is neutral. In this type of fermentation, which is accompanied by the development of an alkaline reaction, and not an acid reaction, the changes are similar to those effected by rennet. It is characterised by the formation of a bitter taste, and by the solution of the coagulum into a somewhat clear liquid, and the formation of butyric acid. It is caused chiefly by the butyric bacillus, but here again we seem to have a number of different bacteria working. Indeed it would seem that there are quite a number of bacteria producing butyric acid.[i] It has been suggested, therefore, to constitute a group of butyric ferments. If, however, as Freudenreich truly points out,[1] butyric acid is rather to be regarded as a residue resulting from the breaking down of

[1] See *Dairy Bacteriology*, p. 88.

casein and milk-sugar in various ways, and is consequently not to be regarded as a uniform process, it is best not to adopt this method of classification.

Casein Ferments.—Just as we have a class of ferments which act upon the milk-sugar, so we have a class that decompose the casein. To the latter class the name casein ferments has been applied. This order of bacteria have also the power of coagulating the milk, not, like the lactic ferments, by the production of lactic acid, but by the production of a rennet-like substance. As belonging to them may be mentioned the *Tyrothrix tenuis*, which has been studied by Duclaux; and a ferment which Conn has succeeded in preparing in the form of a powder, and which acts like rennet ferment. Some of these ferments have the power of redissolving the curdled milk by means of second ferments. They play an important part in the ripening of cheese, and the decomposition of the casein which they initiate is of a most complicated character, and is accompanied by the production of such substances as peptone, leucin, tyrosin, and butyric acid. Among them are the hay and potato bacilli, and a number of so-called butyric ferments. Those best known are those which have been studied by Duclaux, to which he has given the name *Tyrothria*, and of which eight or nine different kinds have been identified. The casein ferments, unlike the lactic ferments, are for the most part spore-forming, and are consequently extremely difficult to destroy, since many of their spores can resist a temperature of about 115° C. (240° Fahr.). It is owing to their presence in it that milk is so difficult to completely sterilise.

There are one or two other forms of fermentation which do not come under any of the classes above mentioned. Among them is *alcoholic* fermentation.

Alcoholic Fermentation.—This kind of fermentation does not often occur in milk; but although it is true that milk does not readily undergo alcoholic fermentation, yet it can be induced[j]; and in certain parts of the world two beverages made from milk which has undergone this kind of fermentation have been long in use. These beverages are *kephir* and *koumiss*.

Kephir.—Kephir has long been used in the Caucasus, and is made from cow's milk as a rule. Investigations were first made into its nature as early as the close of last century, and since then it has been made the subject of further research. It is made by the help of a special ferment known as "kephir grains." This ferment is in the form of hard, yellow, granular lumps about the size of a pea. By soaking these kephir grains in water and then adding them to milk, alcoholic fermentation is speedily induced, and the beverage is ready for use in the course of two or three days. The object of soaking the grains, which has the effect of making them swell, is for the purpose of increasing their activity. After the beverage is prepared, the grains are taken out of it, dried, and kept for future use. When dried thèy may retain their fermenting power for years.

A certain amount of mystery surrounds these kephir grains. Their origin is unknown, since they have been used from time immemorial. Their composition, although investigated at considerable length, is also somewhat of a mystery. They seem to contain a number of different kinds of bacteria and moulds, among which yeast fungi predominate. The other micro-organisms which have been found, according to different investigators, in kephir grains are the *oidium lactis, leptothrix dispora Caucasia* (?), *saccharomyces cerivisiae, bacillus acidi lacti* (lactic bacilli), *bacillus butyricus*. The following chemical analysis of kephir will show its composition (Hammarsten) :—

Water .	88·915
Fat .	3·088
Casein .	2·904
Lactalbumin	·186
Peptone Bodies .	·067
Sugar . . .	2·685
Mineral Salts . .	·708
Alcohol . . .	·720
Lactic Acid . . .	·727
	100·000

Struve found the kephir grains contained 51·69 per cent of albuminoids and 3·99 per cent of fat.

Nature of Kephir Fermentation.—Although the exact nature of the kephir fermentation is not known, there can be little doubt that it is a very complicated one. It is certainly not, as was at first believed, a simple case of the alcoholic fermentation of milk-sugar by means of yeast fungi; although it has been found that certain kinds of yeast, present in kephir grains, are able to produce alcoholic fermentation. The number of fermentative products present in kephir shows that a number of micro-organisms are active in the process, some of them giving rise to alcohol, others to peptone and a variety of acids and other bodies, and others coagulating the casein.

Koumiss.—Although not exactly identical with kephir, koumiss is of a similar nature. In the steppes of Russia, where it has been long used, it is generally prepared, not from cow's milk, but from mare's milk. It is, like kephir, a foaming liquid, resembling butter-milk or sour whey. It differs from kephir slightly in composition, and from the fact that, while kephir may be made direct from cow's milk, koumiss can only be made after the addition of a little cane-sugar. The following is an analysis of koumiss (Fleischmann) :—

		Mare's Milk.
Water	.	91·53
Milk-Sugar	.	1·25
Lactic Acid	.	1·02
Casein	.	1·91
Fat	.	1·27
Alcohol	.	1·85
Carbonic Acid	.	·88
Mineral Matter	.	·29
		100·00

Koumiss may be made by adding a little cane-sugar and yeast to skim milk.

Both these beverages have a high dietetic value, and koumiss has of recent years been used by medical men as a tonic. The dietetic value does not seem to be due to the alcohol they contain, but chiefly to the peptonised condition of the casein, which is thus more easily digested.

Pathogenic Germs.—The bacteria which we have just been discussing may be regarded as, for the most part at any rate, normal inhabitants of milk. There are, however, a class of microbes which may be regarded as abnormal. These are the so-called pathogenic or disease-producing bacteria, which we have already mentioned, but not yet discussed. It may be well to discuss very briefly this question.

The method in which these pathogenic germs act differs in the case of different germs. Some directly absorb their nutriment from the surrounding substance by decomposing it; while some produce *enzymes*, which in their turn cause a chemical decomposition, the products of which are utilised by bacteria as food materials. We have such bodies as *ptomaines* and *toxin* thus produced. Old cheese is on this account rather a dangerous article of food.

Bacillus of Tuberculosis.—The pathogenic microbe which undoubtedly is most commonly present in milk is the bacillus of tuberculosis (see Fig. 13, p. 77). As we have already pointed out, the presence of this bacillus is probably nearly always due to the presence of tuberculosis in the cow. According to Hirschberger, 10 per cent of the cows living in the neighbourhood of towns, where they are not treated properly, suffer from tuberculosis; and 50 per cent of these yield milk containing tubercle bacilli. The virulent character of the bacilli in such milk has been proved by experiment. In Copenhagen four out of twenty-eight samples of mixed milk proved virulent when injected under the skin. Freudenreich[1] quotes a case, cited by Brouardel, where five out of fourteen young girls living together in a boarding-house became consumptive, subsequent to the daily use in the establishment of milk from a tuberculous cow.

It is truly an unfortunate fact that tubercle bacilli seem to be able to live in dairy products for a very long time. In butter they have been found to

[1] *Dairy Bacteriology*, p. 43.

be alive after the lapse of 120 days; and in cheese after the lapse of 35 days. After such an interval of time as a month, they probably become so attenuated as no longer to be very dangerous. It is noteworthy that the tubercle bacillus is not killed at such a low temperature as the majority of bacteria, which, as we have already noted, are killed at a temperature of under 60° C. (140° Fahr.). Experiments have proved that it is necessary to heat milk for thirty minutes at a temperature of 65° C. (149° Fahr.), fifteen minutes at 68° C. (155° Fahr.), and ten minutes at 75° C. (167° Fahr.); but it would be better to boil the milk in cases where the presence of tubercle bacilli is suspected.

Cholera Bacillus.—That milk may be the means of spreading cholera has been proved beyond doubt in several cases. The cholera bacillus (see Fig. 13, p. 77) is not, however, capable of living so long in milk as that of tuberculosis. In ordinary milk, indeed, it seems to be killed in the course of twenty-four hours; in boiled milk, however, it is capable of multiplying abundantly. The reason of this is to be found in the fact that in unboiled milk the lactic acid formed by the numerous lactic bacteria present exercises a speedily fatal effect on the cholera bacilli; in boiled milk, on the other hand, the conditions are more favourable, since the lactic ferments have been killed by the process of boiling. In butter, it would seem that the cholera bacilli may live for four or five days. In cheese, experiments show that they do not seem to be a source of danger. Here again the effect of boiling the milk would be to destroy the cholera germs.

Typhus Bacillus.—Another pathogenic germ which has been found in milk is that giving rise to typhus (see Fig. 13, p. 77). The number of typhus epidemics traced to this source has already been referred to. The bacillus of typhus has been found to be able to survive in butter for a period of from five to seven days.

With regard to the behaviour of other pathogenic germs in milk we know very little. That such diseases as scarlet fever and diphtheria have been propagated by milk we have abundant proof; but as the microbes causing these diseases have not as yet been isolated, we know little about the conditions under which they develop.

As already pointed out, we have certain moulds and yeasts which frequent milk or milk products. Some of these moulds—as, for example, the widely distributed green mould *penicillium glaucum* (see Fig. 10, p. 73)—are familiar in cheese; indeed, the last-named mould plays an important part in the ripening of Roquefort cheese. It is also familiar in Gruyère, Gorgonzola, and Brie cheeses. Another kind of mould common in milk is the *oidium lactis*, abundant in sour milk. It presents a fine white velvety appearance. Among the yeasts may be mentioned the *saccharomyces lactis*, the *saccharomyces acidi lactici*, and the *saccharomyces ruber*.

Methods of Destroying or Regulating Bacterial Life in Milk.—Before concluding this Chapter, it may be well to say a word or two on the methods we have at our disposal for regulating or for entirely checking the development of bacteria in milk.

At first sight the task of excluding bacteria from milk, or even regulating their development, seems to be a hopeless one. But this is not so. Perfect sterilisation is not, under present practical methods, possible, nor, indeed, is it of such great importance, although in many cases highly desirable. In the first place, we cannot expect to entirely prevent the entrance of bacterial life to milk in the process of milking. We may, however, by the exercise of scrupulous care and the observance of cleanliness, minimise contamination. If milk is dirtily and carelessly handled, bacteria, with spores possessing extremely resistant properties, are apt to take possession of it, such as those of the butyric, the hay, and potato bacilli, which render subsequent sterilisation by heat an extremely difficult task.

Sterilisation of Milk.—A distinction should be made between two terms which are often used synonymously, viz. sterilisation and Pasteurisation. Perfect sterilisation of milk can only be effected by submitting the milk to the action of continuous steaming for two hours at a temperature of 120° C. (248° Fahr.), or for thirty minutes at a temperature of 130° C. (266° Fahr.) (Fleischmann). Sterilisation is the term generally applied to the employment of temperatures as high as, or higher than, the boiling point of water. Submitting milk to a high temperature is objectionable, however, for several reasons. When so treated, the composition of the milk undergoes a certain amount of change, which not merely affects its biological condition, but also its physical condition. The soluble lime salts which it contains are converted into an insoluble condition. This change prevents the milk from forming with rennet a cohesive coagulation, the coagulation under such conditions being flocculent. Furthermore, the original fine state of division of the fatty globules is partially destroyed, and a large number of them come to the top. The milk also assumes a dirty brown or yellowish colour, and a strong taste of boiled milk.

Pasteurisation of Milk.—The word Pasteurisation is applied to the use of heat at comparatively low temperatures, those generally ranging from 60° to 64° C. (140° to 147° Fahr.). Heating milk even at comparatively low temperatures, such as in the case of intermittent pasteurisation, effects certain changes in milk, but nothing like to the extent that is done by heating to a temperature of 120° C. (248° Fahr.).

Pasteurisation imparts to milk a slight taint of the cooked flavour, but this flavour is removed on the milk being cooled down. The maximum temperature, therefore, to which milk may be submitted in pasteurising should be below that which imparts to the milk a permanent cooked flavour. Thus it has been found that while heating the milk to a temperature of about 65° C. (149° Fahr.) for twenty minutes induces the flavour of cooked milk, this flavour is not permanent, and quickly disappears on cooling. According to Duclaux, milk suffers permanent change in taste when heated to 70° C. (158° Fahr.). This temperature, then, would seem to be the limit to which milk should be heated in pasteurising. Pasteurisation, however, merely temporarily checks fermentation, since it does not kill all the bacterial spores in the milk. The curdling of milk which has been pasteurised is rather different in its character from ordinary sour milk curdling, and is brought about by bacteria, which are able to excrete rennet. Lactic acid bacteria, as a rule, do not form spores.

For the above reasons intermittent sterilisation, on theoretical grounds, is to be preferred to all other methods. Unfortunately, however, it is such an inconvenient method, and requires so much time, and is so little suited for extended application, that it cannot be carried out on a wholesale scale. At present, therefore, it is impossible to effect the perfect sterilisation of milk. We must be content, accordingly, with partial sterilisation, such as is effected by pasteurisation. Intermittent sterilisation, we may mention, may be carried out by heating the milk for two hours at a temperature of from 70° to 75° C. (158° to 167° Fahr.), then keeping it at a temperature suitable for germ development, viz. about 40° C. (104° Fahr.), in order to permit of the spores which are left behind to develop into adult bacteria. The milk is again submitted for two hours to a similar temperature, and then again allowed to stand for several days at the same favourable temperature, 40° C. (104° Fahr.). These consecutive changes of temperature are repeated four or five times, and at last the milk is brought to 100° C. (212° Fahr.). But, while complete sterilisation of milk may be described as hardly within the range of practical dairying, partial sterilisation is easily effected. It is a well-ascertained fact that bacteria commonly occurring in milk, and which excite its common fermentation, as well as the most dangerous pathogenic germs, can be easily and surely destroyed by heating for an hour at a temperature of from 68° to 75° C. (154° to 167° Fahr.), or by heating for three-quarters of an hour at 100° C. (212° Fahr.) with steam. It is not difficult, therefore, to perfectly sterilise milk which only contains the common bacteria, and at the same time not to affect to any extent its colour, its chemical composition, and the state of division of its fatty globules. Unfortunately, milk of this description—that is, free from spores of a very resistant nature—is of rare occurrence in ordinary practice. As a rule, milk will be found to contain a number of resistant spores, which enter it owing to careless handling.

Unfortunately the *bacterium subtilis* (hay bacillus), a bacterium possessing most resistant spores, is one of the most common in the dust of the byre. Another source whence resistant spores are apt to enter milk is the proximity of fermenting foods kept in the neighbourhood of the byre: such a food, for example, as *silage*. It may be added that the pasteurisation of milk is calculated to do much to reduce the spread of disease among infants, whose sole food is milk. As exemplifying this, it has been claimed that the introduction of pasteurised milk among the poor people of New York has been instrumental in greatly reducing infant mortality during the hot summer months. It is important also to point out that pasteurisation has no injurious effect on milk intended for butter-making purposes. No difficulty is experienced in obtaining the required texture and grain of the butter. The relative digestibility of sterilised as well as pasteurised milk, as compared with ordinary milk, is a question which has been much debated. It seems to be highly probable, however, that neither pasteurisation nor sterilisation of milk impair—to any extent at any rate—its digestibility.

Importance of Cleanliness in Handling Milk.— From what has been above said, it will be seen that the sterilisation of milk is sometimes easily effected, and sometimes with extreme difficulty.

As illustrating the effect which even heating at a comparatively low temperature has on the keeping qualities of milk, it may be mentioned that Fleischmann found by numerous experiments that milk pasteurised at 70° to 75° C. (158° to 167° Fahr.), and then maintained at a temperature of 12° to 14° C. (54° to 57° Fahr.), kept at least thirty hours longer than ordinary milk. If sterilisation is to be effectual, it cannot be too strongly impressed upon the mind of the dairyer that the greatest care should be taken in handling the milk in as cleanly a manner as possible. It is from the impurities which are apt to enter into milk through dirty handling, such as particles of manure, of skin, of food, hair, woollen threads, cobwebs, etc., that the resistant spores common to milk are derived.

As exemplifying the influence of cleanliness, an experiment carried out by the eminent German experimenter Soxhlet may be cited. A cow which had been kept in a badly ventilated byre was milked without having its udder previously cleansed. It was found that under such circumstances the milk, kept at a temperature of 60° Fahr., coagulated in fifty hours. The milk drawn from the same cow, but under different conditions, viz. in an orchard, *i.e.* in the open air, and after the udder and hands of the milker had been cleansed, only coagulated, although kept at the same temperature, after the lapse of eighty-eight hours.

Too great stress cannot be put on the importance of cleanliness, since milk is such an admirable medium for supporting bacterial life. It readily absorbs bad gases, therefore precautions should be taken that the situation of the dairy removes it from any risk of contamination from such sources. The danger of infection of milk from dust may be strikingly illustrated by the statement that 1 gramme ($\frac{1}{28}$th of an ounce) of dust from the road at Turin was found by an Italian investigator, Maggiora, to contain no less than seventy-eight millions of bacteria. The air of rooms in which milk is kept should be cool. If the air is warmer than the milk, the tendency of the milk to absorb impurities therefrom is very much increased. Air in milk-rooms should also be kept dry. Of no less importance is cleanliness in the milk vessels.

It has been pointed out by Hueppe, an eminent bacteriologist who has investigated the subject, that in effecting the sterilisation of milk destined for the use of children—that is, where perfect sterilisation is desirable—the milk should be previously submitted to the action of the centrifugal separator. He asserts that in the slimy residue which is separated from the milk in this way most of the micro-organisms present in the milk, and especially those possessing great resistant properties, are to be found, and may thus be removed from the milk and render its subsequent sterilisation much easier. On the other hand, it has been asserted that in separation by centrifugal force, the cream receives the greatest number, and that the milk becomes much more largely charged with bacterial life, from the air, which is sucked in in such large quantities into the machine when in operation. According to O. Lugger,[1] one litre of milk containing 2,050 millions of germs, on being passed through a centrifugal separator, had these germs distributed in the various products, in the following proportions, viz. 1,700 millions in the cream, 560 millions in the butter-milk, and 18 millions in the dirt. It seems to be true, however, that one disease-producing germ generally separates out in the dirt, viz. the tubercle bacillus. It is not necessary to enter into a discussion of the various methods which have been suggested by Soxhlet, Hueppe, and others, for effecting sterilisation of milk for household purposes, etc.

[1] See *Minnesota Experiment Station Record*, 1893, p. 285.

While the application of heat is the most favourable method of effecting sterilisation, the preservation of milk may also be effected by the application of cold. Cold, however, while it may retard the development of bacteria, does not, unfortunately, entirely destroy them. Certain pathogenic germs have been found capable of resisting the lowest attainable temperatures. If kept in refrigerators, milk remains sweet for a long time. Unfortunately, a serious objection to such treatment of milk is the separation of the cream from the skim milk. A change in the physical condition of the milk seems to be effected, with the result that milk which has been frozen is very different from ordinary milk. It is for this reason that the attempts which have been made, on a large scale, more especially in Paris, have resulted in failure. We may add that cooling milk to a temperature of 10° C. (50° Fahr.), while it does not affect the physical condition of the milk, exercises an important preservative influence.

Chemical Agents.—Various chemical agents have again and again been used as preservatives. Obviously the most effective disinfectants, such as corrosive sublimate, carbolic acid, etc., cannot be used on account of their poisonous nature; whereas such substances as salicylic and boracic acids, carbonate of soda, quicklime, and hydrogen peroxide have to be used in such quantities, if they are to be at all effective, as to render them also unsuited.[k] All authorities, however, unite in condemning the use of preservatives for milk.

([a]) The diseases which may be directly transmitted from the cow to man are—in addition to tuberculosis—scarlet fever, diphtheria, and foot and mouth disease. ([b]) A few dairy bacteria are occasionally known to have pathological characters. Four of these are *Bacillus coli communis*, *Staphylococcus albus*, *Staphylococcus pyogenes aureus*, and *Streptococcus pyogenes aureus*. ([c]) It should be pointed out that the actual number of bacteria in milk is of little significance. The most wide variations have been found in the same milk under different conditions. ([d]) Gessard has quite recently obtained several distinct varieties of this organism by cultivating it under different conditions. These varieties differed widely in the type of pigment they produced. ([e]) Gessard's work, above referred to, seems to indicate that, in order to produce the typical blue colour, the acid must be lactic acid. ([f]) It is somewhat misleading to speak of red milk and violet milk in the same category as blue milk. Blue milk is a dairy infection which attacks the milk rapidly, and produces in a short time a very prominent pigment. In the case of the other types mentioned the pigment is slow in appearing, sometimes a few days, and sometimes several weeks elapsing before it makes its appearance to any appreciable extent. ([g]) Freudenreich has recently disproved this theory of Hueppe's. He has succeeded in separating the bitter-tasting material in a partially free condition, and finds it is not a peptone. ([h]) Up to the present time considerably over 200 dairy bacteria have been described. ([i]) Up to the present time about a dozen organisms are known to produce this kind of fermentation. There are many other kinds of fermentation in which butyric acid is produced as a by-product. ([j]) Alcoholic fermentation of milk is so well recognised now that it has been suggested to use whey for the formation of alcohol, by inoculating it with the proper species of bacteria. ([k]) Recently it has been discovered that potassium bichromate exerts a very important antiseptic influence on milk, from ·3 to ·4 of a gramme per 500 cc. of milk, it is stated, being sufficient. Quite recently also, formalin, viz. a 40 per cent solution of formaldehyde, has been used with great success as a preservative.

CHAPTER VII

Butter—Importance of Bacteria for Butter-Making

BUTTER may be described as the most important of milk products; and although the conditions and nature of the changes going on in the process of the conversion of milk into butter have been made the subject of considerable investigation, there can be no doubt that we are still merely on the threshold of our knowledge of this highly important operation. The nature of the bacteria which effect the many subtle changes going on in butter-making, and the precise nature of the changes they bring about in the butter-fat and other constituents of butter, are questions regarding which we are still very ignorant.

As we have already remarked, in milk, so far as it is used by itself and not as the raw material for the manufacture of butter or cheese, the presence of micro-organic life must be regarded as undesirable. We consequently direct our attention towards destroying, or, as far as possible, limiting, the action of such micro-organic life. In the case of butter and cheese, however, micro-organic life, far from being inimical, is absolutely necessary—indeed, on it depends the whole success of butter and cheese making; and the chief object aimed at by the butter or cheese maker—however unconsciously this object may be pursued—is to produce conditions favourable for the development of the proper kind of micro-organic life.

Object of Churning.—The chief constituent of butter is, as all are aware, butter-fat. But butter-fat is not, as it is too common to suppose, the only constituent. We have in butter, in addition to the butter-fat, all the other constituents of milk—in small amount, it is true, but exercising a not unimportant function in furnishing nutrition for the development of the bacterial life so necessary for the production of good butter. The relative proportions in which these constituents are present are similar to those in which they are present in milk. But, while these minor constituents of milk must not be regarded as of no importance, yet the chief and most important constituent of butter is the butter-fat, and the first great object of the butter-maker is to obtain from his milk the largest possible amount of the butter-fat it contains.

Fat in milk, as we have seen, is in the form of globules. In butter-making these globules are made to coalesce by churning the milk, the fat globules being converted from a liquid condition to a solid one by the violent impact which they thus experience (Figs. 15, 16, 17, and 18, pp. 116-119). Butter can be obtained by directly churning the milk; but it

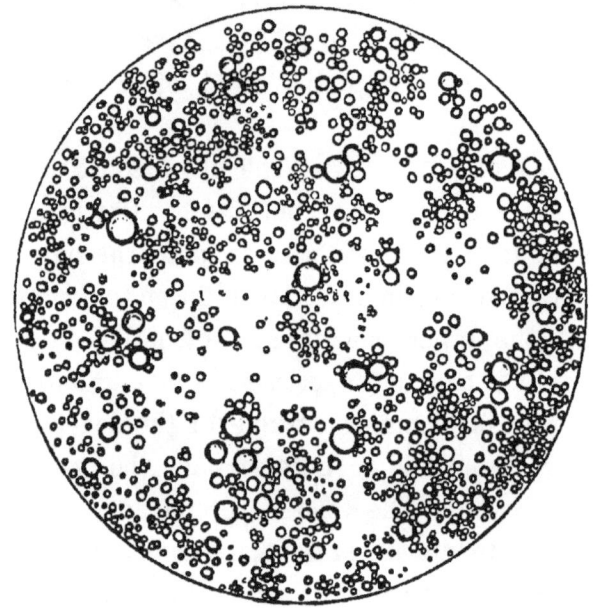

Fig. 15.—Microscopical Appearance of a Drop of Milk, containing 3·6 per cent of fat, magnified 670 times. (Kirchner.)

has always been regarded as more convenient to first cream the milk, and thus effect a concentration of the globules before churning. In the first place, therefore, it may be well to consider what the conditions are which regulate the separation of the fat.

Conditions influencing Separation of Fat.—A

number of influences regulate the rate at which the fatty globules separate themselves from the rest of the milk by rising to the surface. What impedes the separation is the fact that we have not in milk an example of "pure emulsion," *i.e.* a mixture of a fat

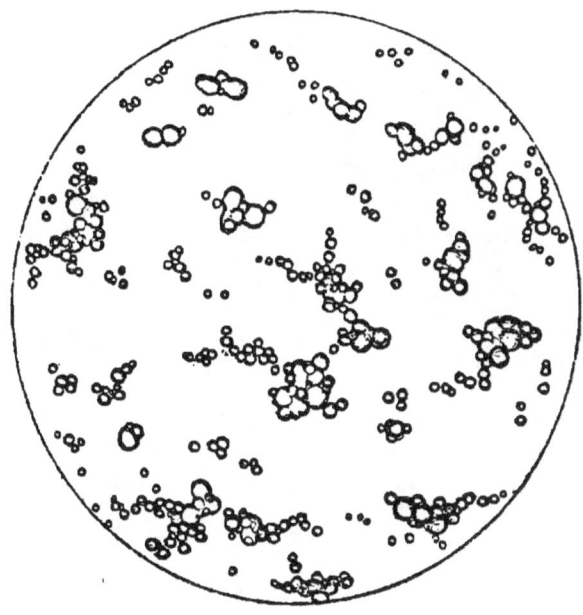

Fig. 16.—Drop of Cream (diluted with water) five minutes after churning has begun. (Kirchner.)

with a liquid. The caseous matter in the milk is in a semi-solid or colloidal form, as was pointed out in a preceding Chapter, and this retards the separation of the fat globules. The temperature of the milk is another important factor. When warm milk is set to cool, currents are set up, and it is difficult to avoid them. The colder portion of the milk, being of greater specific gravity, sinks to the foot; and the warmer portion, being lighter, rises to the top. In this way the collection of fatty globules on the surface is disturbed and impeded. The descending

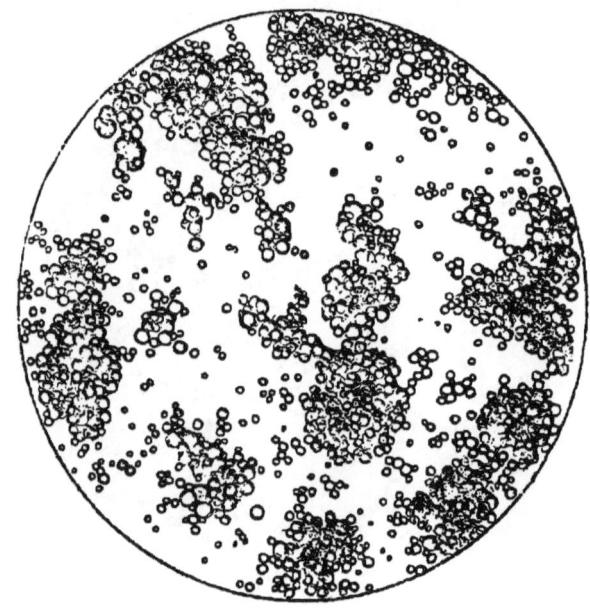

FIG. 17.—DROP OF CREAM (undiluted) fifteen minutes after churning has begun. (Kirchner.)

currents carry away more fat with them from the cream layer than the ascending currents bring back to the surface. It is only after the entire mass of the milk assumes the same temperature as the surrounding air, and when currents due to temperature are no longer induced, that the fatty globules can follow their tendency to collect on the surface without disturbance. The higher the temperature of the milk the less is the resistance offered to the separation of the globules. It may be further pointed out that lactic fermentation is unfavourable for separation.

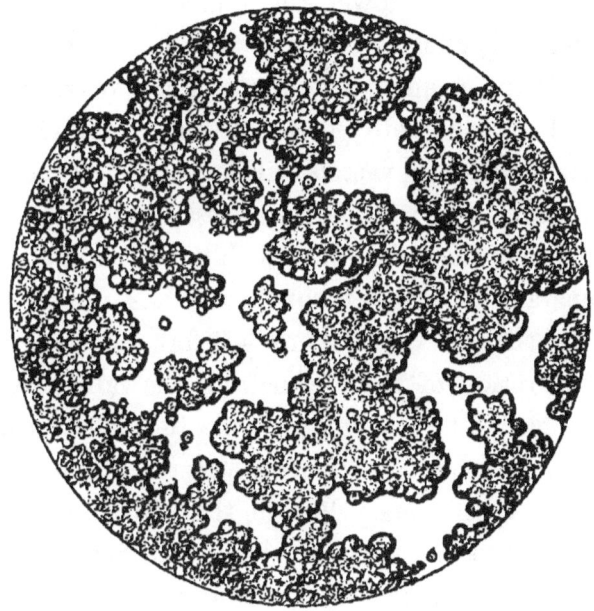

Fig. 18.—Drop of Cream shortly before end of churning.
(Kirchner.)

The sooner, therefore, after milking, the milk is set, the better, since the conditions for its separation are more favourable during the first few hours than subsequently.

The rate at which the globules separate is not equal. During the first hour or two most of the fatty globules come to the surface, and separation thereafter takes place very slowly. The percentage of fat in the cream, *i.e.* the richness of the cream, will be found to vary very considerably, according to the different conditions under which it is formed. Thus, for example, the lower the temperature at which separation is effected the less will be the percentage of fat in the cream. The shape of the vessel, again, also affects the quality of the cream. Milk creamed in narrow-necked vessels yields poorer cream than milk creamed in wide-necked vessels. While, lastly, the depth of the milk from which the cream is allowed to separate influences its quality.

Centrifugal Separators.—Before the year 1877 the separation of the cream from the milk was entirely effected by allowing the milk to stand at rest for a lengthened period—varying, as a rule, from twelve to forty-eight hours. But in the above-mentioned year, a method for effecting this separation, by means of centrifugal force, was introduced, and ever since then the use of "separators"—as the apparatus in which this method is applied are called—has steadily increased. The special advantages of the separator over the older methods are several. For one thing, the saving of time they effect is great, seeing that they effect a much more complete separation of the fatty globules, in about a twentieth or so of the time.

It is impossible to obtain all the fat by creaming, no matter how long the milk may be permitted to stand. There are always a certain number of the smaller fatty globules which never rise to the surface, and which thus remain in the skim milk (Fig. 19).

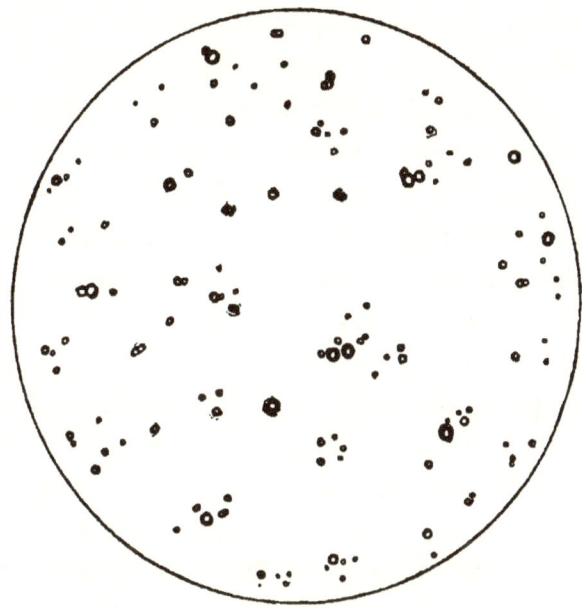

Fig. 19.—Microscopical Appearance of Skim Milk. (Kirchner.)

Even by using centrifugal separators it seems impossible to separate all the fatty globules from the milk. It has been calculated that the amount of fat left in the skim milk by the old methods of separation may amount to, on an average, ·8 per cent. By the use of separators, on the other hand, not more than, on an average, ·2 per cent should be left in the skim milk. From calculations made, it has been estimated that, working on a large scale, the separator yields about 93 per cent of the fat. This is probably about 20 per cent more than is obtained by the old method. The saving effected by using the centrifugal separator is thus seen to be considerable.

The efficiency of the centrifugal separator is dependent on the rate at which it is worked, as well as on the temperature, composition, and amount of milk treated.

Milk transported to a distance, and consequently submitted to a disturbance, or milk which has been boiled, does not separate so well as fresh milk. It has also been noticed that milk very rich in fat does not separate so thoroughly as milk containing an average amount. And here it may be well to point out that skim milk produced by a separator forms a better article of food—when used for the feeding of pigs—than that obtained by the other methods; for, although it contains less fat, it is freer from lactic acid, and forms a much more palatable food.

Preliminary Souring of Milk or Cream before Churning.—It was early discovered that milk or cream, in which the process of souring had gone on to a certain extent, was more easily churned, and yielded more satisfactory results, than when in the fresh condition. It has hence been customary to "ripen" cream before churning. Why this should be so is not, as yet, clearly known. We have referred to Babcock's theory with regard to the formation of a small amount of fibrin in milk. It is possible that the effect of the ripening process is to gradually dissolve the fibrin, and thus permit of more easy coalescence on the part of the fatty globules.

Conditions influencing Churning.—It has been found that temperature is among the important conditions which influence the ease with which coalescence of the fatty globules takes place. It would seem that the obstacles which retard their union decrease with the increase of temperature. Where the process goes on too quickly, however,—and this may be caused by churning at too high a temperature, or by too rapid a motion of the churn,—it has been found that the little lumps of butter do not separate out readily, and are apt to include, in addition to the solidified fat, fatty globules in the liquid condition. The presence of liquid fat in the raw material gives rise to a smeary condition in the butter. It may be mentioned in passing, that there is a slight rise of temperature in the churning operation, amounting to a few degrees. According to Professor Fleischmann, this should not exceed, in the case of churning sour milk or sour cream, two to five degrees Fahr.; and, in the case of sweet cream, six degrees Fahr. According to the same authority, the temperature at which churning should be begun in sweet cream is 12·5° C. (54·5° Fahr.), and for sour cream 16° C. (60·8° Fahr.). Sweet cream seems to require to be churned for a longer time than sour cream, and the loss of fat is very much greater in the former case than in the latter,— indeed, the loss has been calculated by certain American experts to amount to nearly 50 per cent more. The length of time during which churning should last may be said to vary from thirty to forty-five minutes. The effect of diluting cream with water before churning is to increase the period of time required. The loss of fat which, under such circumstances, takes place is slightly greater, and the resulting butter is not so firm.

The Bacteria of Butter. — We now come to consider a very important aspect of butter-making, viz. the relation of bacterial life to butter. It is beyond doubt that the flavour, aroma, and the many qualities which good butter possesses, are due to the action of bacteria. We have already pointed out that, after separating the cream from the milk, it is soured, or "ripened," for a number of hours before being churned. This is effected by allowing it to stand in a warm place for some time, and sometimes by adding some sour cream. Now one reason for this is that the milk globules, under such circumstances, coalesce more easily than they do in sweet cream. But a more important reason is that such treatment permits of the development of certain kinds of bacteria which have a most important effect on the subsequent aroma and flavour of the butter. During this souring period, the initiatory steps of various forms of fermentation are produced.

Aroma and Flavour of Butter.—Butter made from ripened cream has always a superior flavour to that possessed by butter made from unripened cream. Indeed, the obtaining of the proper flavour and aroma are among the chief objects aimed at by the butter-maker in butter-making. How the aroma in butter is exactly produced is not very clear. It has been thought that it is due to some alcoholic-like product formed during ripening, or, it may be, to the presence of lactic acid itself; but, so far as this latter statement is concerned, lactic acid used artificially for ripening cream has not been found very successful. The flavour is probably due to the presence of certain volatile acids in the butter which are not present in fresh milk; but, whatever the cause, there can be little doubt that both aroma and flavour are connected with decomposition products formed by bacterial growth.

The enormous importance of having the proper sort of bacteria developed will be at once seen from what has been above stated. The experiments, therefore, which have been carried out with a view to isolating the bacteria implicated are of the very highest importance. The practical outcome of these experiments has been the preparation of so-called "pure cultures" for ripening cream. So far the use of these pure cultures has been followed with the very best results, and it has been found that the quality of the butter improves as soon as these are used. According to experts, the amount of the improvement has been estimated at more than 20 per cent. At present it may be mentioned that nearly every dairy in Denmark, Germany, Sweden, and Holland, depends upon pure cultures of bacteria to produce uniformly good qualities of butter, and it is greatly to be desired that this practice should be introduced into this country without loss of time. Very favourable results have been found by using such cultures for correcting certain defects in butter, such as *oiliness, fishy* and *bitter* flavours, and a tendency to become *rancid*. One bacillus has been isolated which seems to produce the proper aroma in butter, when used in cultures for ripening cream, and this ferment has been used in some of the creameries of Germany with excellent results.[1] It may be interesting to note that the proper aroma in butter seems to be connected with the first products of decomposition, set up in the cream as the result of bacterial growth; for the bacteria which

[1] See Appendix to Chapter, p. 136.

have been found growing in ripening cream may produce all sorts of disagreeable flavours. While pleasant flavours seem to belong to the first products of decomposition, those formed during the later stages give rise to disagreeable ones.

Since, then, the great danger in ripening is to continue it too long, and thus induce a disagreeable flavour in the butter, a point of first-rate practical importance is to determine the proper length of time for ripening. But this is just exactly what it is very difficult to do. The butter-maker can have no certainty that his cream is supplied with the proper species of bacteria; hence the importance of being able to furnish to butter-makers the proper ferment, and thus avoid all such risk. Investigation into this subject is but in its initial stages, and only one species of bacteria has as yet been discovered which produces the proper results. It is along this line of investigation that the most useful results from future researches in the science of butter-making are to be looked for. We are the more encouraged to expect this on the analogy of the great benefits which have accrued to the brewing industry from the application of similar principles. The great uncertainty which accompanied the manufacture of beer in former times has been greatly reduced by the use of pure yeast cultures. As illustrating the intimate connection between bacterial life and flavour in butter, it may be mentioned that a species of bacteria has been found which produces an odour and taste similar to that produced by the feeding of roots. Butter made by the use of this kind of bacteria acquired this root taste in about two weeks.

Number of Bacteria in Butter.— Interesting investigations on the number of bacteria present in butter have been recently carried out by Lafar on different samples of Munich butter. In one gramme (about the $\frac{1}{28}$th of an ounce), taken from the centre of a pat of butter, nearly two and a half millions were found. Equal amounts of butter, taken from more exposed portions of samples, were found to contain far larger numbers—in one case as many as forty-seven and a quarter millions being found. Taking the average of a number of investigations, Lafar concluded that a gramme of butter contains from ten to twenty million bacteria.

The few experiments which have been carried out on the subject show that artificial butter contains fewer bacteria, and those of a different nature, than natural butter. The presence of pathogenic bacteria in butter is dangerous, for the reason that butter cannot be submitted to sterilisation, but has to be eaten in the raw state. Lafar is of the opinion that typhus and cholera bacteria would die in butter in from five to seven days, and those causing tuberculosis lose their power of doing harm in twelve days.

Influence of Lactation, Breed, Age, and Food on Quality of Butter.

—Among the other conditions which influence the quality of butter may be mentioned the *lactation period*, the *breed*, and the *age* of the cow. Butter made from the milk of cows which have been milked for some time is generally firmer than that made from the milk of newly milked cows, and possesses a less fine flavour. With regard to the influence of feeding on the condition of the butter, it has been proved that both the colour, the flavour, the smell, the keeping qualities, and, in a very special degree, the solidity of the butter-fat, are dependent on the nature of the food used. Especially intimate seems to be the relation between the food and the fatty acids. The volatile fatty acids are at their maximum with a new period of lactation and at the beginning of the pasturage season, and decrease with the advance of the season and period of lactation. There seems to be no connection between the fat in the food and the fat in the butter. From extensive experiments carried out on the effect of food on butter, Professor Adolf Mayer has come to the conclusion that rations rich in carbohydrates have a favourable effect in increasing the volatile fatty acids. It must be frankly admitted, however, that the evidence on the effect of different foods on the composition of butter is very conflicting. Some interesting experiments, having for their object the comparison of the daily yield of butter obtained from 1 lb. of fat in the milk of different breeds of cows, including Jersey, Guernsey, Holstein (German), Ayrshire, Devon, and Holderness (American), showed the following results:—

	lb. Butter.		lb. Butter.
(1) Guernsey	1·07	(4) Devon	·97
(2) Jersey	1·04	(5) Ayrshire	·93
(3) Holderness	·98	(6) Holstein	·88

It may be added that the size of the globules seems to affect the quality of the butter, since the composition of large fatty globules seems to differ from the composition of small ones (see Milk-Fat, Chapter II., p. 9). Butter made from large globules has a richer colour, a better taste and consistency, a lower melting-point and point of crystallisation, contains less insoluble fatty acids, and, finally, possesses a lower specific gravity than butter made from smaller globules.

Chemical Composition of Butter.—In conclusion, a word may be said on the chemical composition of butter. A point of considerable importance, and one regarding which there has been much discussion during recent years, is the amount of water which butter should contain. On the Continent, the concensus of opinion seems to be that the limit should be fixed at 15 per cent, while in this country and in America a larger limit seems to be supported. A large number of analyses of American butter showed an average of about 19 per cent of water. The question is of considerable importance, since the amount of water undoubtedly affects the quality. The percentage of fat may be said to vary from 82 to 85 per cent, the other constituents—the casein, albumin, and ash—forming, on an average, not much more than 1 per cent. For preserving purposes salt is often added to butter.

The following may be regarded as the average composition of fresh and salt butter:—

	Fresh Butter.	Salt Butter.
Water	14·00	12·50
Fat	83·50	84·50
Protein	·80	·50
Milk, Sugar, etc.	1·50	·60
Ash	·20	·10
Salt	...	1·80
	100·00	100·00

The following diagram illustrates the appearance of butter under the microscope (see Fig. 20, p. 132).

Margarine.—Before concluding this Chapter, it may not be out of place to say a word or two on that largely used butter substitute, *oleomargarine*, or, as it is more commonly known, *margarine*. Having regard to the present enormous dimensions of the trade in margarine, it is striking to reflect that its manufacture only dates from the year 1870. It was in that year, a few months before the outbreak of the great Franco-German war, that the Emperor Napoleon III. commissioned M. Mège-Mouriès to carry out experiments with a view to discover a

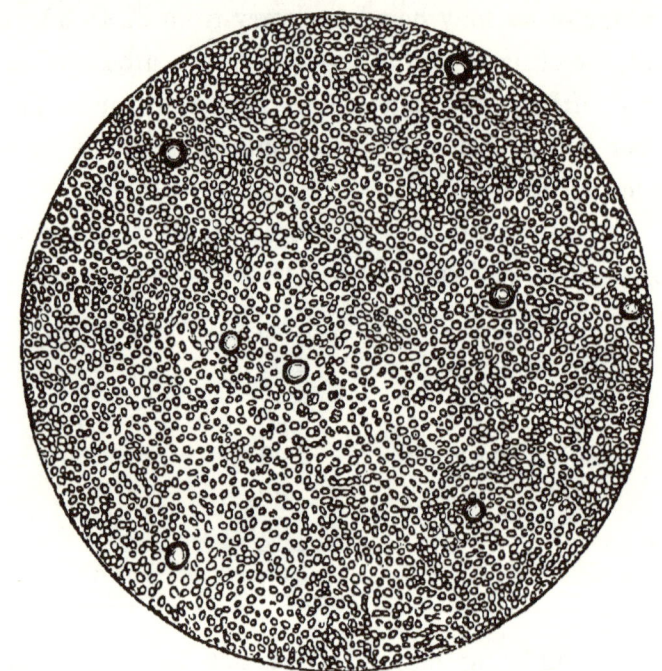

Fig. 20.—Butter. Magnified 350 diameters. (Bell.)

good substitute for butter, for the use, in the first instance, of the French marines and the poorer inhabitants of Paris. The result was an article which was first known as "Margarine-Mouriès." This substance was prepared in a simple manner from the best animal tallow.

The preparation of the new butter substitute rapidly extended and became established in America, Russia, Germany, Austria, and other countries. Up to the end of the year 1880 nearly all the margarine sold in Europe was prepared according to the French process; and as the new fat was found to be an excellent cooking fat, more economical for cooking purposes than butter itself, since it contained a larger percentage of fat, and as it also kept better, and was undoubtedly both in quality and flavour superior to the inferior kinds of butter, its use and popularity steadily increased. The large extension in its manufacture, however, had the result that the raw material, viz. fresh ox tallow, first exclusively used in its preparation was soon no longer procurable in the necessary quantities. Other oils, chiefly the cheaper plant oils, such as cotton-seed, earth-nut, walnut, cocoa-nut, sesame, and the poorer sorts of olive oil, in addition to ox tallow, bacon fat, goose fat, fat from soap-boiling manufactories and from slaughter-houses, etc., had to be used. Further, the mode of preparation was considerably changed, and the fat, in the process of extraction, was submitted to a higher temperature and a greater pressure than in the original process, with the result that an inferior article in every way was prepared. Margarine, therefore, from being an admirably cheap butter substitute possessing many qualities to recommend it, soon became, under the altered conditions of manufacture, an article to be viewed with suspicion. This suspicion was further increased by the reflection that the larger part of the raw material from which it was manufactured in the margarine factories was imported from America and Australia, viz. from sources not under control. In this way, it is not impossible that certain kinds of infectious diseases may be propagated by its use; and badly prepared margarine may possibly contain the spores of animal parasites originally present in the fat constituting the raw material, and may thus be introduced into the human system. Since, in the preparation of the article, only a comparatively low temperature can be used, these organisms may very easily escape destruction. Lastly, among the least serious objections which can be urged against the use of margarine made from plant fats is the fact that they are less easily digested than animal fats.

There can be no doubt that during the early years of the trade much of the margarine made was far from a desirable article; and while the manufacturers of butter might naturally feel aggrieved at having, as a rival of butter, an article which, in many cases was of an inferior nature, yet, so long as the article was sold in a fairly legitimate manner, they had no recourse but to put up with it. The case was altered, however, when this same article was fraudulently sold in the place of butter, and largely mixed with pure butter. There can be no doubt that the practice—at first, it is possible, innocently followed—of producing in margarine an article which resembled genuine butter in flavour and external appearance, as well as in the form in which it is packed and exposed for sale, led the way to fraudulent practices. The very name, butterine, under which it was at first permitted to be sold, had a fraudulent ring about it.

The extent to which this fraudulent sale of margarine was practised gave rise in the year 1885 to an attempt being made, in most countries where the dairying industry was in an advanced state, to deal with this practice. The result of the movement was that laws were passed in most European countries forbidding the sale of the article under the name of butterine, and enforcing other restrictions. There can be no doubt that this legislation has done much to lessen the fraud; yet it is equally certain that an enormous amount of adulteration of butter with margarine still goes on.

The chief difference, from a chemical point of view, between margarine and butter is that the former contains a larger percentage of fixed insoluble fatty acids than the latter, which may be said never to exceed 90 per cent of the total quantity of fat. The specific gravity of the fat of margarine is also lower than that of genuine butter fat. The following diagram illustrates the appearance of margarine under the microscope (see Fig. 21).

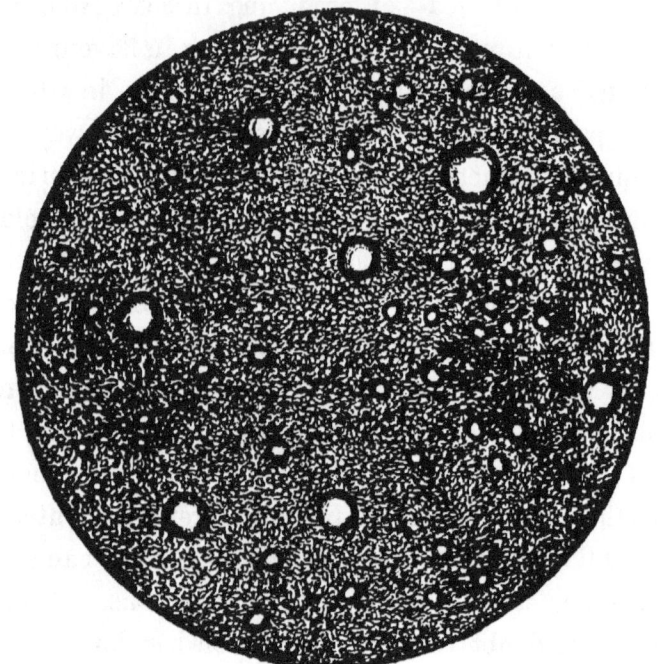

FIG. 21.—OLEOMARGARINE. Magnified 350 diameters. (Bell.)

APPENDIX

Conn points out that an objection to the pure cultures at present in use on the Continent is the fact that they consist of acid-producing bacteria, and that when added to cream containing lactic ferments they unduly hasten fermentation. The result is that cream, before being treated with them, has to be pasteurised. This is objectionable, since it imparts to the butter a "cooked" flavour. He claims that by the use of an organism, which he calls Bacillus No. 41, this is obviated. His bacteria, instead of hastening the souring of the cream, actually retards it, so that it is not necessary to pasteurise the cream before adding it.

CHAPTER VIII

Rennet and Its Action

BEFORE passing on to consider cheese, it will be best to discuss the nature and action of rennet.

The chief constituent of ordinary cheese is the casein of milk, and the object of the cheese-manufacturer, therefore, is to separate this casein from the milk. This is effected by coagulating it. From a very remote period it has been customary to effect the coagulation of milk in either of the two following ways, viz. by allowing it to become spontaneously sour, or by treating it with rennet. For long these two processes were regarded as essentially the same in their nature, and although several observers noticed that certain differences existed—in the reaction produced, for example, by the respective kinds of coagulation—little notice was paid to this fact.

Occurrence of Rennet Ferment.—Before discussing the action of rennet, it will be well to say a few words on the nature of this substance. With regard to this, there is little precise information. It is, as all are aware, a preparation usually made from the stomach (the rennet stomach) of the calf; but it is also found in the stomach of the sheep, and, although in smaller quantity, in the stomachs of many other animals, the deer and the chamois, as well as in fishes and birds. Whether it is always of identically the same composition when obtained from these different sources or not, it is impossible to say. So much, at any rate, may be said, that it is possible to gain from the stomachs of the above-mentioned animals, on treating them with salt or lactic acid, a substance which exerts an action on milk similar to that exerted by rennet from the calf's stomach. Rennet has also been found in the human stomach. It is a secretion of certain glands embedded in the lining of the stomach, and is most abundant in the very young animal, especially during the period of suckling. As the young animal, however, gradually ceases to depend for its food on milk, the production of rennet decreases. It has also been found in a number of plants. For example, the juices of the fig tree, the artichoke, certain kinds of thistle, and the melon tree, as well as other plants, have been found to contain a substance, the action of which is similar to that of rennet. Whether, however, this substance is identical in all cases, and whether it is the same as the rennet obtained from the calf's stomach, is also unknown. Finally, it may be mentioned that rennet, or a rennet-acting substance, is produced by numerous bacteria.

Active Principle of Rennet.—While it has been found impossible hitherto to isolate rennet in an absolutely pure condition, Hammarsten (to whose researches is due most of our information on the nature of rennet, as well as its action) has obtained, in a comparatively pure state, what he regards as its active principle. To this he has given the name *lab*, while the terms *chymosin* and *pixine* have also been applied to it by other investigators. It belongs to that class of ferments which are known as *chemical* or *unorganised* ferments (also known as *enzymes*), and which have only been distinguished from the organised ferments for about thirty years. To the same class belong *diastase*, a substance which converts starch into sugar, and *pepsin*, which dissolves the albuminoids. Both rennet and pepsin are found in the gastric juices of the stomach, and both are to be regarded as digestive ferments of the highest importance. These enzymes, or chemical ferments, are produced by the growth of the organised ferments. These two classes of ferments—the true or living ferment, and chemical ferment or enzyme—are distinguished from one another by certain properties. The former are insoluble in water, while the latter are soluble. Again, while the chemical ferment can only effect a certain amount of fermentative work, which can be definitely measured, there is no limit to the amount of work the organised ferment may effect. As already pointed out, a most important relationship exists between these two ferments; for most true ferments give rise to a chemical ferment, which is one of the products of their action. It is on this account often very difficult to know whether certain fermentative changes are caused by the true or the chemical ferment.

Coagulating Power of Rennet.—As has been pointed out, the amount of work which unorganised ferments can effect is capable of being definitely measured. Now some conception of the coagulating power of rennet is afforded by the following short description of an experiment carried out by Söldner, a German investigator. He prepared an extract of rennet by treating the dry stomach of a calf with a 5 per cent solution of salt. The rennet was then precipitated, in the form of a grayish-brown powder, by adding more salt. He found that one part of this powder in the dry state was able to coagulate, in forty minutes, at a temperature of 35° C. (95° Fahr.), one million parts of milk. On analysis, he found that this brown powder only contained 36 per cent of organic matter; and as this 36 per cent was probably not the pure rennet, the inference is that one part of pure rennet can coagulate, at the above temperature and in the above period of time, more than three million parts of milk.

Difference between Rennet-Curd and Acid-Curd.—The nature of the action of rennet is a point regarding which our knowledge is still very imperfect. If the coagulation of curd produced respectively by the formation of lactic acid or other acid (and this curd may be distinguished by calling it "acid-curd," and that produced by rennet "rennet-curd") are examined, it will be found that these two curds differ in many respects in their properties and composition. Rennet-curd, for example, is an elastic substance scarcely soluble in water, and not in the slightest extent sticky or greasy. Acid-curd, on the other hand, is not elastic, is less insoluble in water, and is sticky and greasy. Again, acid-curd, will be found to exhibit a strongly acid reaction; while, although the reaction in rennet-curd is also acid, it is only slightly so. It is for this reason that acid-curd is not so suited for the manufacture of such a wide variety of cheese as rennet-curd, since its acid reaction is not favourable to the development of a large variety of micro-organic life. On the other hand, in rennet-curd the conditions for bacterial development are much more favourable. Hence in cheeses made from acid-curd, with very few exceptions, the process of ripening resembles in general the process of putrefaction, inasmuch as it goes on from outside to inside.

In cheese made from rennet-curd the ripening process goes on throughout the whole mass. Furthermore in rennet-curd the calcium phosphates which are in suspension in the milk are entirely enclosed in it, whereas acid-curd, it will be found, only encloses a small quantity of these phosphates, and this for the reason that most of them are dissolved by the lactic acid. Lastly, acid-curd contains more fat than rennet-curd.

Nature of Action of Rennet.—Now, if we examine into the nature of the rennet coagulation we shall find first of all, as Hammarsten found in his researches, that it is entirely independent of the formation of an acid in the milk, and that the reaction of the curd is not changed during the curdling. Although the curd does, as a rule, exhibit an acid reaction, this is due to the action of micro-organisms. Hammarsten, secondly, proved that the action of rennet is entirely independent of milk-sugar. This he proved by curdling solutions of casein, which had been entirely freed from sugar, with rennet. Now these two points alone are sufficient to distinguish the acid coagulation of milk from the rennet coagulation; for it is known that the acid coagulation of milk, as effected in milk which is allowed to spontaneously coagulate, is due to the action of true or living ferments, to which the name lactic ferments has been applied, which act upon the milk-sugar, decomposing it into lactic acid. This lactic acid coagulates the milk when it reaches a certain amount. Acid coagulation of this type, therefore, may be said to be entirely dependent on the presence of milk-sugar.

That the curds respectively produced by rennet and by acid are very different, Hammarsten further proved by the following experiment. He took some curd which had been produced by acid coagulation, and, dissolving it in a little alkali and then neutralising the solution, he obtained a solution of casein, which he found he could not coagulate with rennet. From this he concluded that there was something present in the original milk which was a necessary condition of the action of the rennet, but which was not contained in the acid-curd. This something, he argued, must be present in the whey which he had separated from the acid-curd. To see if this was so, he added to his casein solution some of this whey and then tested it with rennet, when he found that it was coagulable by rennet. Further investigation demonstrated that this something, which was lacking in the acid-curd, was lime salts. He thus discovered the further interesting fact that rennet coagulation requires the presence of lime salts, and found that other alkali salts can take the place of lime. In fact, he came to the conclusion that the curdling by rennet is similar to the clotting of blood.

It has also been found that milk possessing an

alkaline reaction is not capable of being coagulated by rennet. A very slight acid reaction, on the other hand, assists its action, which is largely influenced by temperature, the most favourable temperature being 30° C. (86° Fahr.). A temperature of 60° C. (140° Fahr.) destroys its action. Milk which has been boiled generally loses its power of being precipitated by rennet, but such milk regains its susceptibility at once if calcium chloride or other soluble lime salt be added, or if the lime salts precipitated by boiling be again dissolved by the addition of a dilute acid.

It has been sometimes noticed that even fresh milk is not coagulated by rennet. Investigation would probably show that such milk, owing to disturbance in the milk-gland, exhibits a slightly alkaline reaction, and does not contain soluble lime salts.

It has already been stated that casein in milk exists in a semi-dissolved or colloidal form. This semi-soluble condition of casein is due to the fact that it is combined in some way with lime. According to this view, the casein present in milk may be regarded as of the nature of a salt, in which the casein proper takes the part of the acid, and the lime that of the base. The coagulation which results on the addition of an acid may thus be explained by the decomposition of this body, the lime being neutralised by the acid and the casein set free in the form of an insoluble clot. Be this as it may, the casein in the milk, to which the term caseinogen has been given, to distinguish it from the casein in the curd, seems to be kept in its state of semi-solution by the alkaline condition of the milk, seeing that the addition of a small quantity of acid at once effects precipitation.

The action of rennet, however, is much more complicated, as it breaks up the casein into two albuminoid bodies, one of which is easily coagulated by acids, and at a temperature of from 70° to 80° C. (158° to 176° Fahr.); whereas the other is not precipitated at all by acids, and is only coagulated by a temperature of from 95° to 100° C. (203° to 212° Fahr.). Now it is only the first of these two bodies that is recovered in the curd, since it forms along with alkaline earths a solid body. The second is in solution, and remains in the whey, and is thus lost. According to this theory, then, rennet does not directly possess the power of precipitating the casein in the milk. All that it does is to decompose the casein into two proteid bodies, one of which has the power of becoming coagulated by the calcium salts, which are always present in the milk. This reaction throws some light on the amount of time required for the coagulation of milk by rennet. It may be pointed out, that the amount of albuminoid matter, lost in the coagulation of milk, is further increased by the fact, that certain bacteria have the power of acting upon the curd and dissolving it; hence it is advisable to use rennet in such a manner that it will produce its coagulation as quickly as possible.

Form in which Rennet is used.—Formerly a solution of rennet was always used. This was of home manufacture, and was made in small quantities for immediate use. Now, however, rennet is largely manufactured on the commercial scale, either in the form of a solution or in a powder. These solutions generally contain, in addition to the active principle of the rennet, small quantities of pepsin, an unorganised ferment which produces lactic acid, and comparatively large quantities of slimy matter and other organic matter, the composition of which is unknown. They also contain salt or alcohol, and sometimes other preservatives, such as boracic acid, salicylic acid, benzoic acid, etc. All these substances increase the keeping qualities of the rennet solution, but this is effected at the expense of its strength, since they render a portion of the rennet ferment inactive.

Extracts of rennet should be clear in appearance, and should not possess a disagreeable smell. Their coagulating properties should not lose, in the course of a year, by more than 25 per cent. Rennet powders are more concentrated than rennet solutions; they should be almost entirely white, and devoid of smell, and should almost completely dissolve in water. Care should be taken, in keeping them, to preserve them from dampness, and when used they must be dissolved in water as thoroughly as possible. Rennet solutions should be kept from the action of light, as this tends to weaken their strength.

CHAPTER IX

Cheese

CHEESE-MAKING may be described as a very ancient art. Fresh cheeses, that is to say, non-ripened cheeses, were probably made thousands of years ago; and although the art of cheese-making was doubtless practised in a more or less crude form, much attention seems to have been devoted from very early times to the different methods of preparing cheese. Thus, for example, in Aristotle's writings we find references to the use of different kinds of rennet.

Important Rôle of Bacteria in Cheese-Making.—Cheese-making may be described as a much more difficult art than butter-making, the number of conditions which have to be reckoned with in it being very much greater than in the case of butter-making. The importance of bacteria in the manufacture of cheese is, again, very much greater than in the manufacture of butter. In the latter case they are highly desirable allies, and, for a satisfactory article, no doubt necessary, since without their help butter tastes flat. Yet butter made without their aid is still capable of being used. On the other hand, in cheese-making, bacteria are absolutely indispensable, since there can be no doubt whatever that what imparts to a cheese its characteristic properties, and indeed renders it so desirable an article of food, viz. its flavour, is due to the action of the different kinds of bacteria or classes of bacteria. The proper flavour of cheese is the result of a ripening process in which the chief agents are bacteria. This ripening process goes on at different rates in different cheeses, and may last for weeks or months. To the cheese-maker, therefore, the assistance which the bacteriologist is likely to afford is of the very highest importance.

Conditions determining Quality of Cheese.—As fat is the chief constituent of butter, so casein is the chief constituent of cheese; but while fat predominates in butter over all other ingredients, in many cheeses the amount of fat present may be quite equal to, and, indeed, largely exceed, the amount of casein. Although this is the case, it is the changes which go on in the ripening process in the casein that are of chief importance. The means taken to produce different kinds of cheeses are various. The nature of the cheese will be determined by a number of conditions, among which may be mentioned the method in which the milk has been coagulated. This determines, to a certain extent, the composition of the curd, its physical condition, such as its fineness. Among other conditions are the degree of moisture in the curd, the percentage of fat, the regulation of the temperature during the numerous processes, and the sourness of the curd. One point of enormous importance in cheese-making is to obtain a similarity of composition in the curd.

Bacteria in Cheese.—That the ripening of cheese is due to the action of micro-organisms has been conclusively proved. Thus, for example, if means be taken to prevent the growth of bacteria, by subjecting the cheese, as has been done, to a stream of carbonic acid gas, or if cheese be made from sterilised cream, it is found that no ripening takes place. Similar results have been obtained by treating the cheese with some disinfecting agent, which would prevent bacterial growth in the cheese without affecting its chemical condition. As to the nature of these organisms, however, and as to the specific functions they exercise, much doubt yet remains. Different investigators have attempted to estimate their number. Thus Adametz found, in a gramme of fresh cheese, from 90,000 to 140,000, and Freudenreich as many as 1,800,000. The former observer has found 850,000 in a gramme of Swiss cheese, and in the same amount, taken from the outer layer, of soft house cheese, 5,600,000. Their number has been found to increase slowly during the ripening process. That these bacteria belong to various species seems to be evident. It seems also pretty conclusively proved that every kind of cheese has to be ripened by a definite species; but the difficulty is in identifying the specific kinds of ferments connected with the different kinds of cheeses, that perfect ripening requires the united action of many different species of bacteria, the action of which upon one another is of a very complicated nature. Certain cheeses, again, not only require various kinds of bacteria in their preparation, but also certain moulds. Some of them are aerobic, while others are anaerobic. How far the characteristic properties of different kinds of cheese are respectively due to the action of one specific class of bacteria is not absolutely certain. There seems to be strong evidence, however, that this is the case; since, as the ripening process goes on, one species generally increases at the expense of others. As the bacteria concerned in the ripening of cheese are always present in milk, the art of cheese-making consists in producing circumstances which will favour the predominant development, of that class of bacteria, specially implicated in the ripening of the particular kind of cheese to be made. A ferment, made by the bacteria which ripen cheese, has been isolated and studied by Duclaux: he calls it *casease*.

Chemical Changes Cheese undergoes during the Ripening Process.—But what, it may be asked, are the chemical changes which cheese undergoes in the ripening process? Ripening, from a chemical point of view, may be said to consist chiefly in the transformation of the characteristic constituent of cheese, viz. the casein, from the insoluble condition in which it is present in fresh cheese into certain soluble albuminoid bodies. The unpalatable nature of fresh cheese is due to the fact that the casein is in an insoluble condition. By allowing the cheese to ripen, therefore, a certain amount of decomposition of the casein is effected, and the solubility of the cheese in water is consequently largely increased. It is, therefore, to the presence of these soluble bodies that the flavour, quality, and condition of a cheese are largely due. What these decomposition products exactly are is still doubtful. We know, however, that the milk-sugar, which is present to a small extent in fresh cheese, is no longer to be found in ripened cheese. It has been decomposed into such bodies as lactic acid and, it may be, butyric acid. The fat in cheese seems, however, to undergo very little decomposition in the ripening process. Among other decomposition products, alcohol, oxalic acid, carbonate of ammonium, leucin, and tyrosin have been identified. During ripening, too, there is generally produced a certain amount of gases, consisting of carbonic acid, hydrogen, and sulphuretted hydrogen. Indeed, the production of gas often gives rise to a not uncommonly occurring "fault" in cheeses, viz. an inflated or puffy condition. This is often caused by premature ripening of the cheese; but the presence of pores in certain kinds of cheese, amounting sometimes to the size of little cavities, is a desirable property. One investigator, who has carried out researches on the swelling of cheese, has come to the conclusion that the formation of holes or pores in hard cheese is due simply to the action of a single bacillus, the *bacillus diatrypeticus casei*. Indeed, the general process of cheese-ripening seems to be similar in its nature to the process of digestion by the various digestive fluids of the stomach and alimentary canal. It may be added that the percentage of water in the ripening process becomes distinctly less. A fact of considerable interest, and one which has an important practical bearing, is the inimical influence which light has on the development of the ripening bacteria.

Cheese-Faults.—While the proper ripening of cheese is thus undoubtedly due to the action of bacterial life, so also are most of the faults which cheeses are liable to. Among these faults may be mentioned fungoid growths on the surface of the cheese, red, blue, or yellow in colour, sometimes leading to the total discoloration of the cheese, as in the case of "black" cheese. Again, the bitter flavour which characterises certain cheeses is undoubtedly due to bacterial life. These faults have been generally attributed in the past to the health and condition of the cow, its food, the condition of the byre, and other causes. These may, no doubt, be the indirect cause, but the direct cause is bacterial life.

The difficulties with which the cheese-maker has to contend, therefore, are great. He cannot always be sure of a uniform product; for, in spite of all precautions he may adopt, disastrous results may be obtained. To bacterial life also is due the poisonous properties which, unfortunately, cheeses have been found not infrequently to develop. These poisonous properties are due to the production of *ptomaines*, which are apt to occur in old cheese.

Imperfect and unsatisfactory as our knowledge at present is of the relation of bacteriology both to butter and cheese making, in this department of research lies the hope of the future of the dairy industry. We have pointed out how very uncertain must necessarily be the results obtained in butter and cheese making under present conditions. This is due to the fact that we have to work so much in the dark in these processes, and thereby incur great risks. While bacteria are the useful servants of the butter and cheese manufacturer, they can also prove his most deadly foes. Unfortunately, it is not always in his power to decide which rôle they shall perform. When, however, our knowledge of the bacteria implicated in these processes has become sufficiently exact to admit of our isolating them and preparing them in pure cultures, we shall then be in a position both to make butter and cheese from fresh milk which has had little time to develop any undesirable kind of fermentation, or from sterilised milk, by simply adding the pure culture of the proper bacteria, and thus secure uniformly good results.[1]

Composition of Cheese.—Cheese, as we have already said, varies very much in composition. Sometimes it contains but a small proportion of fat, while in other cheeses the fat is the most abundant constituent. There is nearly always present a small quantity of lactic acid. Good cheeses contain from 30 to 35 per cent of water; inferior kinds from 38 to 45 per cent. Rich cheeses contain from 25 to 35 per cent of fat, sometimes even more. The casein is very variable in amount. In skim-milk cheese it amounts to as much as 50 per cent; while in some of the soft cheeses it may be as low as 11 to 12 per cent. The following table illustrates the composition of the commoner kinds of cheese [2]:—

[1] That there is great room for improvement in the manufacture of cheese may be inferred from the statement that, according to Dr. Conn, sometimes as high as 50 per cent of the cheeses made in a factory are worthless, or comparatively worthless, owing to abnormal ripening.

[2] Johnston's *Elements of Agricultural Chemistry*, Seventeenth Edition, by C. M. Aikman, p. 461.

Soft Cheeses.

	Water.	Casein.	Fat.	Ash.
Neufchâtel	37·87	17·43	41·30	3·40
Fromage de Brie	51·87	18·30	24·83	5·00
English Cream	...	2·30	50·68	...
Camembert	51·30	19·00	21·50	4·70
Roquefort	11·84	85·43	1·85	...
,, (two months old)	19·30	43·28	32·30	...

Hard Cheeses.

	Water.	Casein.	Fat.	Ash.
American	22·59	37·20	35·41	4·80
,,	31·80	36·00	28·70	3·50
Cheddar	27·83	44·47	24·04	3·66
,,	28·34	47·03	21·01	3·62
Dunlop	38·46	25·87	31·86	3·81
Gloucester	21·41	49·12	25·38	4·09
,,	35·82	37·96	21·97	4·25
Parmesan	27·56	44·08	15·95	5·72
Stilton	32·18	24·31	37·36	3·93
,,	38·28	23·93	30·59	3·20
Gruyère	40·00	31·25	24·00	3·00
,,	34·68	31·41	28·93	3·85
Gorgonzola	43·56	24·17	27·95	4·32
Skim	43·14	49·79	0·86	6·21

CHAPTER X

Testing of Milk

A SHORT description of the methods in use for testing milk may be given. To do so, various methods may be employed. These may be divided into two classes : (1) tests and (2) chemical analysis, the former of which depend chiefly on the physical properties of milk. Taking the former class of tests first, these are as follow :—

1. *Determination of the percentage of fat by measuring the thickness of the layer of cream thrown up on allowing it to stand for some time.*

2. *A determination of the fat by optical tests.*

3. *Determination of the fat by the amount of butter the milk yields.*

4. *Determination of the fat by the addition of certain reagents, etc.*

5. *Testing the milk by determining its specific gravity.*

1. **Testing Fat by Amount of Cream.** — With

regard to the determination of the percentage of fat in milk by the thickness of the cream layer, it must at once be said that, although this method has long been popular on account of its simplicity, it is a most unreliable one. An apparatus which is largely used for this purpose is the *Chevalier Cremometer*, which consists of a glass cylinder of about 20 centimetres deep and about 4 centimetres broad, possessing a graduated scale which indicates the percentage of fat corresponding to different layers of cream. The milk is allowed to stand in this for twenty-four hours, and the depth of the cream layer then read off. The reason of the inaccuracy of such a method of testing the fat in milk is due to the fact that, as has been pointed out elsewhere, the same quantity of fat in milk will, under different conditions, give layers of cream of very varying depth. It may be said that ordinary milk will give, on an average, about 10 per cent of cream; but so variable is this that it may sometimes amount to from 30 to 40 per cent, and at other times may be as low as from 4 to 5 per cent, and this despite the fact that the range in the amount of fat present may be very trifling. Experiments carried out by Kirchner exemplify this. In four different samples of milk in which the percentage of fat only varied to the extent of ·03 per cent, a difference in the percentage of cream, amounting to 4 per cent, was found, owing to the difference in the conditions under which the tests were made.

2. **Optical Method of Determination of Fat.**—A variety of instruments have been devised for applying this test. It is based on the fact that the opacity of milk is dependent on the milk-fat globules. As, however, this is only partly the case, and as the same weight of fat retards more light when it is in the form of small globules than when it is in the form of larger globules, this method of testing is very unreliable.

3. **Determination of the Fat by Churning.**—Apparatus has also been devised for the determination of the fat in milk by the amount of butter it yields on churning. This method, since it involves the use of very expensive apparatus, is not suited for general use.

4. **Determination of the Fat by the Addition of Reagents.**—Of the various methods for the determination of the fat by the addition of reagents, Soxhlet's is the best known. This method consists in dissolving the fat from a measured quantity of milk, to which some potash or soda has been previously added, by ether, and calculating its amount from the quantity of fat dissolved in the layer of ether which separates out, from the specific gravity of the latter. The apparatus by which this test is carried out is known as the '*areometer*.'

Another admirable method for determining fat is by the *lactocrit*, an apparatus devised by Dr. de Laval. A measured quantity of milk is placed in a glass tube, an equal amount of concentrated acetic acid, containing 5 per cent of concentrated sulphuric acid, is then added, and the mixture heated for a few minutes at boiling temperature. The effect of the acid is to dissolve the caseous matter, which retards the movement of the fatty globules. The tube is then placed in a specially constructed centrifugal apparatus, and the mixture is separated by centrifugal force, the depth of the layer forming a measure of its amount.

Another method is that devised by Leffmann and Beam. A quantity of milk is treated in a bottle with a graduated neck with one-fifth of its volume of a mixture of equal parts of amyl alcohol and strong hydrochloric acid. Strong sulphuric acid is then added to the mixture in nearly equal quantity, and the whole is well mixed. The bottle is then placed in a centrifugal separator, which is rotated for one or two minutes. The fat layer, which collects in the neck of the bottle, is then read off.

5. **Determination of the Specific Gravity.**—The determination of the specific gravity of milk must be reckoned as one of the most important of all the tests to which it can be subjected. Alone, it is no doubt apt to be misinterpreted. In milk we have

not, as in a mixture of water and alcohol, merely two factors determining specific gravity. In milk we have *water, fat*, and *solids not fat*. The tendency of the fat, since it possesses a lower specific gravity than water, viz. ·93, is to counteract the tendency of the "solids not fat," which possess a higher specific gravity, viz. 1·6. Two interpretations, therefore, of a low specific gravity in a sample of milk are possible, viz. the addition of water, or excessive richness in fat. It is for this reason that the determination of the specific gravity alone must not be overrated in value.

Formulae for Calculating Composition of Milk from certain Data.—In conjunction, however, with a knowledge of the total solids of milk, the specific gravity furnishes a most useful indication of the percentage of fat; and similarly, if the fat be known, it permits us to calculate the percentage of total solids. This can be effected by using the following formulae (Fleischmann), in which

T = percentage of total solids in milk.
F = percentage of fat in milk.
G = specific gravity at 15° C.

(*a*) For calculating the percentage of fat from the specific gravity and total solids:—

$$F = \cdot 833T - 2 \cdot 22 \frac{100G - 100}{G}$$

(*b*) For calculating the percentage of total solids from the fat and specific gravity :—

$$T = 1\cdot 2F + 2\cdot 665\frac{100G - 100}{G}$$

The following is a formula which has been devised by Hehner and Richmond, and which has been largely used in England :—

$$F = \cdot 859T - \cdot 2186G$$

This formula suffices for normal milk, but for skim milk it has been found necessary to modify it as follows :—

$$F = \cdot 859T - \cdot 2186G - \cdot 05\left(\frac{G}{T} - 2\cdot 5\right)$$

The above correction has only to be applied when the specific gravity (G) divided by total solids exceeds 2·5.

H. Droop Richmond has quite recently[1] devised a formula for calculation of total solids from the specific gravity and fat. It is as follows :—

$$T = \frac{G}{4} + \frac{6}{5}F + \cdot 14$$

Chemical Analysis of Milk.—With regard to the chemical analysis of milk, the subject is one which does not admit of being discussed with any fulness in a work like the present. It may suffice to mention

[1] *Analyst* (March 1895), p. 57.

that, as ordinarily carried out, it includes a determination of the total solids, the fat, and the ash. The water and the solids not fat are calculated by difference. The total solids are estimated by evaporating at boiling temperature, on the water bath, a weighed quantity of milk to dryness. The percentage of ash is obtained by incinerating at a low red heat. For the extraction of fat various methods are in use. The most satisfactory one undoubtedly is that known as Adam's coil process, which consists in absorbing the milk by means of a coil of filter-paper, which is then subsequently dried and transferred to a Soxhlet extraction apparatus, where the fat is dissolved out by ether and weighed.

CHAPTER XI

Milk as a Food

THE position which milk occupies as a food is unique. That it acts in the capacity of a perfect or complete food is well known to all, since it is the first and often the only food of the young. It may not be, therefore, without interest to say a word or two in conclusion on its value in this connection, so far as the subject has been investigated. In order to do so it will be necessary to state shortly the principles upon which the value of a food rests.

Classification of Food Nutrients. — All food materials, whether animal or vegetable, generally consist of two portions—the edible and the refuse. The refuse portion consists of such parts as the bones of meat and fish, the bran of wheat, and the skin of potatoes. The edible portion, on the other hand, may be divided up into a number of groups of substances of different chemical composition, of varying digestibility, and charged with different functions. To the members of each of these different groups the name nutrient is given. Of these nutrients there are a large number; but they can all be classified under a few groups, this classification being based upon their chemical composition. Thus we have the *protein* group, the characteristic feature of which is the possession of nitrogen in its composition; the *carbohydrates*, of which starch and sugar are typical members; and the *fats*. The first group may be subdivided into several smaller groups, such as *albuminoids, gelatinoids, amides,* and *nitrogenous extractive substances*. Lastly, we have another class of nutrients, which are also, in their way, of importance, viz. *mineral nutrients*. Now, while all of these groups of different nutrients perform important functions, they cannot be described as of equal importance.

Functions of Food.—The functions of food are various. Inasmuch as the body is constantly experiencing change, is always undergoing, as we say, a certain amount of "wear and tear," and is also, at certain periods of life, viz. in youth, undergoing a process of growth, one obvious function of food is to make good this wear and tear, and to build up the tissue, bones, etc., out of which the body is formed. Since animal life, in building up its tissue, is only able to make use of foods more or less similar in their nature to its tissue, it follows that the muscles, tendons, and other portions of the body which contain nitrogen in their composition can only be built up by members of the protein group of nutrients. But, as has been pointed out, the protein group is a large one, and all its members do not possess this tissue-forming power; indeed, it is only the albuminoids—so called from the typical member of the group, albumin, or the white of an egg—which possess it.

The albuminoids, then, are the most important group of nutrients. From them also is the most of the fat in the animal body formed; although both the fats and the carbohydrates are also capable of being used in this way. The proportion in which the fat in the body is derived from the three sources is not known, and, indeed, varies with circumstances, more especially according to the proportion in which the three different classes of nutrients are present in the food. As the body, more especially in certain of its parts, such as the bones, blood, teeth, hair, and many of the fluids, contains mineral salts (phosphates, sulphates, and the chlorides of potassium, sodium, calcium, and magnesium), certain mineral substances are absolutely necessary.

But another great function of food is to furnish the body with the necessary animal heat. Although the members of the protein group are capable of acting in this way as heat-givers, and furnishing the necessary fuel to keep the animal machine working, this is more especially the function of the other two groups of nutrients, viz. the fat and the carbohydrates. But since the carbohydrates are unable to discharge any functions as flesh-formers, and can form fat probably to only a very limited extent, their characteristic function may be described as heat-givers.

And here it may be well to point out with regard to the value of the three members of the three groups of nutrients as heat-givers, that they are not all equal in this respect, and that fats, when burned in the body, yield $2\frac{1}{2}$ times as much heat as an equal weight of a protein or carbohydrate nutrient. A common method of comparing different foods is to compare their heat-giving value. Another method of comparing foods is by calculating what has been named their *albuminoid* or *nutritive* ratio, viz. the ratio which their flesh-forming nutrients bear to their heat-giving nutrients. The amount of albuminoids or flesh-forming nutrients in almost all foods is very much less than the amount of fat and carbohydrates. It is customary, therefore, in calculating the nutritive ratio, to take the amount of albuminoids as 1. To obtain the other number of the ratio, the amount of fat present in the food is first multiplied by $2\frac{1}{2}$, and then added to the weight of the carbohydrates. Thus, supposing that a food contains per pound, 2 ounces of albuminoids, 2 ounces of fat, and 2 ounces of carbohydrates, the nutritive ratio of such a food would be

$$2 : 2 \times 2\tfrac{1}{2} + 2 = 7, \text{ or } 1 \text{ to } 3\tfrac{1}{2}.$$

Digestibility of Food Nutrients.—But the value of a food as a source of nourishment does not solely depend on its composition, viz. the percentage of albuminoids and other nutrients it contains, but also on the extent to which these nutrients are capable of being digested. And here we are met with a difficulty, since the digestibility of a food nutrient cannot be absolutely stated, as much depends on the digestion of the individual partaking of the food. Indeed, no better example of this fact could be cited than the case of milk, which is generally found to be one of the most digestible of foods, yet which some people find to be very indigestible. What the cause of such an anomaly is, it is difficult as yet to say. It may be, as has been suggested by recent researches on this subject, that ferments in the digestive canal of some people may cause particular compounds to be changed into injurious, and even poisonous forms, so that it sometimes may be literally true that "one man's meat is another man's poison." But digestion proper is a chemical process, and is capable of being determined with fair accuracy by careful experiment, and our knowledge on this subject has received many valuable additions in the last few years. These have shown that the digestibility of the different nutrients in different foods varies very considerably, and that the feeding value of a food should never be solely estimated by simply taking into account the percentages of the different nutrients it contains.

Value of Milk as a Food.—If we examine the composition of milk, in the light of the above necessarily brief statement of the nature and functions of the different food nutrients, we shall find why it should occupy such a unique position among foods. In the first place, it contains no refuse portion; and, in the second place, it contains members of all the different groups of nutrients. The albuminoids are represented by *casein* and *lactalbumin*, which amount to, on an average, $3\frac{1}{2}$ per cent; the fats are represented by *butter-fat*, which amounts to $3\frac{3}{4}$ per cent; while the carbohydrate group is represented by the *milk-sugar*, averaging about $4\frac{1}{2}$ per cent; and, lastly, we have in the *ash* the necessary mineral ingredients.

Digestibility of Milk.—With regard to the digestibility of these food nutrients in milk, broadly speaking, it may be said that, so far as the small number of experiments, carried out on this subject, go to show, all the protein and all the carbohydrates are digested. The same, however, is not the case with the fat, which, it would seem, is only digested to the extent of about 96 per cent. In cheese, again, the same may be said, the fat being slightly less digestible, viz. to the extent of 95 per cent. In butter, however, it is probably the same as in milk. While lastly, in margarine, the fat is less easily digested, viz. only to the extent of 95 per cent. Of course this refers to the case of adults endowed with sound digestion. As it has been already pointed out, milk in many cases may prove far from an easily digested food. But while a nutrient may be perfectly digested, the time required for this operation differs in the case of different foods. Thus, in certain experiments carried out to show the relative digestibility of meat, mutton, milk, etc., in the uncooked and cooked state, the following results were obtained :—

	Hours necessary for Digestion.
Beef—	
Raw	2
Boiled, half done	$2\frac{1}{2}$
Boiled, well done	3
Roasted, half done	3
Roasted, well done	4
Mutton—	
Raw	2
Veal—	
Raw	$2\frac{1}{2}$
Pork—	
Raw	3
Milk—	
Uncooked	$3\frac{1}{2}$
Boiled	4
Sour	3
Skimmed	$3\frac{1}{2}$

Effect of Boiling on Milk.—We have already pointed out the result the heating of milk has on its nature. We saw that among the changes which took place was the coagulation of the albumin, which forms a skin on the surface of the milk. Now, while such a change would not affect the ultimate digestibility of the albumin, it would no doubt retard it. This doubtless accounts for the fact that, as the above table shows, milk in the uncooked state is more speedily digested than in the boiled condition, as indeed is seen to be the case with the other foods cited. The difference, however, is slight. Again, what may strike the reader as, at first sight, singular, is that sour milk is more speedily digested than either fresh or boiled milk. This is probably due to the fact that the lactic acid generated in the souring process helps the gastric juices in the stomach in their solvent action on the different nutrients.

Suitability of Milk as a Food.—If it be asked whether milk contains the different food nutrients in the best possible proportions for sustaining animal life, we may safely answer that it does, so far as children are concerned. Certain foods we know are better adapted for the digestive organs of children than adults, and this is the case with milk. Again, it has been found that the composition of the mineral constituents of milk is very similar to the composition of the mineral matter in the body of the sucking animal, with slight exceptions. In one respect there is a strange anomaly, and this is that the percentage of iron in the latter is about six times that in the former. This has been explained on the ground that the amount of iron in the body of the young is relatively much greater than in the body of the adult; that, in short, iron is stored up in the body of the young for future use. Milk has also been found to be richer in potash and poorer in soda than the body of the sucking animal. Another point which renders milk less suitable for adults than for young is its extremely bulky nature, and the fact that it contains an excess of fat. A greater proportion of water and fat is required in the food of the young than in the food of the adult. It has been calculated that, if one were to live on a milk diet alone, eleven pints per diem would be required to be consumed in order to afford proper nourishment.

Comparing it with such a food as lean beef, it may be said that 1 lb. of beef contains about the same quantity of actually nutritious materials as a quart of milk. The food which comes nearest milk in the amount of nutriment it contains is oysters, which are practically of the same nutritive value as milk. There can be no doubt, however, that while milk is perhaps not suited to act as the sole food of the adult, it is one of the best and cheapest articles of diet we possess, and should be far more widely used than at present is the case. It is one of the most convenient, useful, and inexpensive sources of albuminoids which we have. Indeed, skim milk may be safely asserted to be the cheapest of all foods.

Value of Butter and Cheese as Food.—With regard to the value of milk products as articles of diet, as far as their heating power goes, butter and cheese stand at the head of all foods. Indeed, butter has twice as much heating power, weight for weight, as any other food except pork and cream cheese; while skim-milk cheese, so far as flesh-forming constituents are concerned, is the most concentrated of all the common foods.

Milk as a Food for Invalids.—While the subject hardly admits of discussion in this place, a single word may be said in conclusion on milk as a food for invalids. Milk cures are of very old date, and have been practised for many years, more especially in Switzerland, where they are carried on, on a large scale, in establishments specially equipped for the purpose. A large and wide experience has shown the great value of milk as a diet for consumptive patients, especially when living at mountain health-resorts. Again, in such diseases as ulceration and catarrh of the stomach it is found in many cases to be the sole possible food. Special preparations of milk have long been used. We have already referred to the use of koumiss in this connection, a beverage made by the alcoholic fermentation of milk. Butter-milk, again, has been found to be of great use for diabetic patients.

APPENDIX

WORKS ON DAIRYING

The Science and Practice of Dairying, by Professor Fleischmann. Translated and edited by Dr. C. M. Aikman and Professor R. P. Wright. (London and Glasgow: Blackie, 1895.)

Principes de Laitière, by E. Duclaux. (Armand, Colin, et Cie., 1893.)

Dairy Industry and Dairy Farming in Denmark. R. P. Ward. (Crewe, 1893.)

British Dairying, by Professor J. P. Sheldon. (London: Crosby, Lockwood, and Co., 1893.)

Dairy Farming, by Professor J. P. Sheldon. (London: Bell and Son, 1893.)

Handbuch der Milchwirthschaft, by Dr. W. Kirchner. (Berlin: Parey, 1891.)

Die Milchdrüsen der Kuh, by Dr. M. N. F. Fürstenberg. (Leipzig: Engelmenn, 1868.)

The Fermentation of Milk, by Dr. H. W. Conn. (United States Department of Agriculture, Bulletin No. 9, 1892.)

A Review of Recent Work on Dairying, by E. W. Allen. (United States Department of Agriculture, *Experiment Station Record*, vol. 5, 1894, No. 10.)

The Souring of Milk. (United States Department of Agriculture, Farmers' Bulletin No. 29, 1895.)

Bacteria in their Relation to the Dairy. (*Annual Report 1893, Agricultural Experiment Station, University of Minnesota.*)

The Principles of Modern Dairy Practice, by Grotenfelt. Translated by F. W. Woll. (New York: John Wiley and Sons, 1894.)

Bacteria in their Relation to the Dairy, by Freudenreich. Translated by Professor Davis. (London: Methuen, 1895.)

Dairy Bacteriology. (United States Department of Agriculture, Bulletin No. 25, 1895.)

Die Neueren Massen-Fettbestimmungsverfahren für Milch, by Dr. P. Vieth. (Bremen: M. Heinsius Nachfolger, 1895.)

Elements of Dairy Farming, by Professor J. Long. (London, Glasgow, and Edinburgh: W. Collins, Sons, and Co., Limited, 1894.)

Theory, Practice, and Methods of Dairy Farming, by Professor J. P. Sheldon. (London: Cassell and Co., Limited, 1891.)

INDEX

ACID-CURD, 141
Aerobies, 80
Age of cows, influence of, on butter, 129; on milk, 45
Albumin, 27, 30
Albuminoids, 165, 166; of milk, 27-31
Albuminose, 31
Alcoholic fermentation of milk, 97
Algae, 71
Alveoli, 3, 4
American cheese, composition of, 156
American Holderness cows' milk, 47
Amides, 165
Amphoteric reaction of milk, 36, 57
Aroma of butter, 125
Average composition of milk, 12
Ayrshire cows' milk, 46, 47

BACILLI, 72, 73
Bacillus, actinobacter du lait visqueux, 88; actinobacter polymorphus, 89; cyanogenus, 85; lactis pituitosi, 89; leuconostoc mesenteroides, 88; mesentericus, 88; prodigiosus, 83; synxanthus, 87; violaceus, 87; viscosus, 89
Bacteria of milk, 63-113
Bacterium Hessii, 89
Beastings, 38-41
Bitter milk, 89

Blue milk, 84
Breeds of cows, milk from different, 45
Budding of yeasts, 80
Butin, 24
Butter, 114-136; aroma and flavour of, 125; bacteria of, 124; chemical composition of, 130; number of bacteria in, 128
Butyric acid, 26
Butyric bacilli, 83, 95
Butyric fermentation of milk, 95
Butyrin, 24

CAMEMBERT cheese, composition of, 156
Capillary blood-vessels, 7
Caprin, 24
Capronin, 24
Caprylin, 24
Carbohydrates, 165
Casease, 151
Casein, 27, 28
Casein ferments, 91, 96
Caseinogen, 27
Centrifugal separators, 120
Changes milk undergoes, 53-62
Cheddar cheese, composition of, 156
Cheese, 148-156; bacteria in, 150; composition of, 155; conditions determining quality of, 149; ripening of, 152
Cheese-faults, 153

Chemical analysis of milk, 162
Chlorestin, 35
Chlorine in milk, 34
Cholera propagated by milk, 68, 76
Churning, 115-124
Chymosin, 139
Cilia, 72
Citric acid in milk, 35
Cleanliness in relation to milk, 109
Coagulation of milk, 57, 137-147
Cocci, 71
Colostrum, 7, 38-41
Colostrum corpuscles, 7, 38
Comma bacilli, 72, 73
Condensed milk, 61
Conditions influencing composition of milk, 42-52
Constituents of milk, 9, 18-41
Consumption propagated by milk, 68, 75
Corps granuleux, 38
Cream, microscopical appearance of, 117-120
Creaming of milk, 54

DEVON cows' milk, 47
Diastase, 139
Digestibility of foods, 168 ; of milk, 169
Diphtheria propagated by milk, 68, 69, 75
Diplococci, 71, 73
Dunlop cheese, composition of, 156

ENGLISH cream cheese, composition of, 156
Enzymes, 101, 139
Erysipelas propagated by milk, 75
Excitement, influence of, on milk, 51

FAT in milk, 18-27 ; condition of, 21 ; conditions influencing separation of, 116 ; determination of, 157-160

Fatty acids, 24
Fatty globules, 18 ; number of in milk, 20
Faults of milk, 59
Ferric oxide in milk, 34
Fibrin, 31
Fission of bacteria, 72, 73
Food, influence of, on butter, 129 ; on milk, 49 ; milk as a, 164-173 ; nutrients, 164
Formation of milk, 6
Formic acid, 27
Formulae for calculating composition of milk, 161
Fromage de Brie cheese, composition of, 156

GALACTINE, 31, 33
Gases in milk, 36
Gelatinoids, 165
Glanders propagated by milk, 76
Gloucestershire cheese, composition of, 156
Glycerides, 24
Goats' milk, 37
Gorgonzola cheese, composition of, 156
Grape-sugar, 33, 40
Green milk, 87
Gruyère cheese, composition of, 156
Guernsey cows' milk, 46, 47

HEAT, effect of, on milk, 58
Holstein Friesian cows' milk, 47
Hyphae, 80
Hypoxanthin, 35

ILLNESS, influence of, on milk, 52
Infusoria, 71
Insoluble fatty acids, 25
Intermittent sterilisation, 61, 106
Invalids, milk as a food for, 173

JERSEY cows' milk, 46, 47

KEPHIR, 97-99
Koumiss, 99
Kreatin, 35

INDEX

Lab, 139
Lactalbumin, 30
Lactation period, 48, 129
Lactic bacteria, 55, 91-95
Lactocaramel, 33, 59
Lactocrit, 160
Lactoglobulin, 31
Lactoprotein, 27, 31
Lactose, 40
Laurin, 24
Lecithin, 35
Leffmann-Beam milk-tester, 160
Leprosy propagated by milk, 76
Leucin, 35
Lime in ash of milk, 34
Lobes, 3
Lobules, 3, 4
Lockjaw propagated by milk, 75

Macrococci, 71
Magnesia in ash of milk, 34
Malaria propagated by milk, 76
Mammary gland, 3, 5
Mares' milk, 37
Margarine, 131-136
Micrococci, 71, 73, 74
Micrococcus prodigiosus, 86; Freudenreichii, 89
Microscopical appearance of milk, 19, 116
Milk as a food, 164-173; bacteria of, 63-113; causes and conditions influencing quality and quantity of, 42-52; changes undergone by, 53-62; constituents of, 18-41; percentage composition of, 9-17; testing of, 157-163
Milk-cisterns, 3
Milk-fat, 18-27
Milk-sugar, 31-33
Mineral constituents of milk, 34-36
Moulds, 72, 73
Myristin, 24

Neufchâtel cheese, composition of, 156

Nuclein, 28

Olein, 24
Optical method for determining fat, 159

Palmitin, 24
Parmesan cheese, composition of, 156
Pasteurisation of milk, 105-109
Pathogenic germs, 68, 75-77, 100-104
Penicillium glaucum, 73
Pepsin, 139
Peptones, 30
Percentage, composition of milk, 9-17
Phosphoric acid in milk, 34
Pixine, 139
Pneumonia propagated by milk, 75
Potash in milk, 34
Potato bacilli, 84, 89
Preservatives for milk, 60, 112
Preserved milk, 61
Preventing changes in milk, 60
Protein group, 165
Ptomaines, 76, 101
Pure cultures of bacteria, 126

Rancid butter, 126
Red milk, 86
Rennet, 137-147; active principle of, 139; coagulating power of, 140; forms in which used, 146; occurrence of, 137
Rennet-curd, 141
Ropy milk, 88
Roquefort cheese, composition of, 156

Sale of Foods and Drugs Act, 14
Sarcina rosea bacillus, 86
Scarlet fever propagated by milk, 69
Secretion of milk, 6-8, 42
Sheep's milk, 37
Shorthorn cows' milk, 46
Skim milk, 121

Skim-milk cheese, composition of, 156
Slimy milk, 88
Society of Public Analysts, standard, 14
Soda in milk, 34
Soluble fatty acids, 25
Souring of milk, 55, 122
Specific gravity of milk, 37, 160
Spirillá, 72, 73, 77
Spores, 73, 74
Standards for milk, 17
Staphylococci, 71, 73
Stearin, 24
Sterile milk, 65
Sterilisation of milk, 61, 104
Stilton cheese, composition of, 156
Streptococci, 71, 73
Streptococcus Hollandicus, 89
Structure of cow's udder, 1-8
Sulphuric acid in milk, 34
Superfusion, 23

Symbiosis, 85

TATTEMYELK, 88
Teats, 2, 5
Testing of milk, 157-163
Toxin, 101
Trimethylamin, 87
Tuberculosis bacillus, 68, 77, 101
Tunica propria, 4
Typhus propagated by milk, 68, 69, 75, 103
Tyrosin, 35
Tyrothrix, 84

UDDER, structure of, 1-8
Urea, 35

VARIATION in composition of milk, 12
Violet milk, 87

YEASTS, 72, 73
Yellow milk, 87

THE END

Printed by R. & R. CLARK, LIMITED, *Edinburgh.*

www.ingramcontent.com/pod-product-compliance
Lightning Source LLC
Chambersburg PA
CBHW020842160426
43192CB00007B/745